Studies in Systems, Decision and Control

Volume 288

Series Editor

Janusz Kacprzyk, Systems Research Institute, Polish Academy of Sciences, Warsaw, Poland

The series "Studies in Systems, Decision and Control" (SSDC) covers both new developments and advances, as well as the state of the art, in the various areas of broadly perceived systems, decision making and control–quickly, up to date and with a high quality. The intent is to cover the theory, applications, and perspectives on the state of the art and future developments relevant to systems, decision making, control, complex processes and related areas, as embedded in the fields of engineering, computer science, physics, economics, social and life sciences, as well as the paradigms and methodologies behind them. The series contains monographs, textbooks, lecture notes and edited volumes in systems, decision making and control spanning the areas of Cyber-Physical Systems, Autonomous Systems, Sensor Networks, Control Systems, Energy Systems, Automotive Systems, Biological Systems, Vehicular Networking and Connected Vehicles, Aerospace Systems, Automation, Manufacturing, Smart Grids, Nonlinear Systems, Power Systems, Robotics, Social Systems, Economic Systems and other. Of particular value to both the contributors and the readership are the short publication timeframe and the world-wide distribution and exposure which enable both a wide and rapid dissemination of research output.

** Indexing: The books of this series are submitted to ISI, SCOPUS, DBLP, Ulrichs, MathSciNet, Current Mathematical Publications, Mathematical Reviews, Zentralblatt Math: MetaPress and Springerlink.

More information about this series at http://www.springer.com/series/13304

Domenico Marino · Melchiorre A. Monaca
Editors

Economic and Policy Implications of Artificial Intelligence

Springer

Editors
Domenico Marino
Mediterranea University of Reggio Calabria
Reggio Calabria, Italy

Melchiorre A. Monaca
Mediterranea University of Reggio Calabria
Reggio Calabria, Italy

ISSN 2198-4182 ISSN 2198-4190 (electronic)
Studies in Systems, Decision and Control
ISBN 978-3-030-45339-8 ISBN 978-3-030-45340-4 (eBook)
https://doi.org/10.1007/978-3-030-45340-4

This Springer imprint is published by the registered company Springer Nature Switzerland AG
The registered company address is: Gewerbestrasse 11, 6330 Cham, Switzerland

Introduction

On 27 and 28 September 2019, the International Seminar on *Economic and Policy Implications of Artificial Intelligence* was held in Reggio Calabria, a welcoming Mediterranean city in Italy.

The development of artificial intelligence will have important repercussions on the economic system and economic policies, opening new scenarios, imposing new constraints, creating new potential risks and developing new opportunities. Artificial intelligence produces and will produce very strong changes in the immediate future. What we are experiencing is a new Industrial Revolution, only much faster than the previous ones, therefore without the possibility of foreseeing medium–long adjustment times.

The approach that guided the design and development of the aimed seminar tried not to emphasize the positive aspects and not to hide the AI risks. As in any social phenomenon, the governance of processes and the definition of a system of rules and policies are the aspects that can create advantages or disadvantages, and above all, they can determine who are the beneficiaries of the advantages and who must bear the disadvantages.

The involvement of professors and researchers from Italian and foreign universities allowed us to develop and enrich scientific knowledge on artificial intelligence applied to the following fields: economics, finance, law, operational techniques, management, production, human resources and business organization.

The level of some of the works presented and discussed in the forum led us to consider the possibility of compiling them in a document, so that they could be published by a world-renowned publishing house. The intense work of the Profs. Domenico Marino and Melchiorre A. Monaca has made possible that today this book becomes a reality.

This is a multi-author book, but not a collection of essays. In fact, although signed by different authors, all the chapters of this book follow a line of development which is traced in the first part of the book and deepen the various aspects in a logical order.

Various parts of the book will explore the most important features of AI and analyse the implications of AI in economics, law, decision-making and smart citizens.

Domenico Marino
Melchiorre A. Monaca

Contents

AI and Smart Citizenship

AI and Economics

A Regional Indicator to Describe Potential Artificial Intelligence

Domenico Marino and Domenico Tebala

Abstract Artificial Intelligence (from now on AI) is synonymous with innovation. In the laboratories of companies and universities, we start to use more innovative technologies to answer specific questions, but in Italy IA is not very widespread. As Gerhard Friedrichn, professor of computer science, dean of the Faculty of Technical Sciences of the University of Klagenfurt, one of the pioneers of research about AI in Europe, says, "artificial intelligence will not come in the coming years, but it will eventually come. There is nothing that can prevent it." Therefore, for the purpose of a better understanding of the phenomenon to date, this study analyzes the potential use of artificial intelligence in the Italian regions through a composite index (using the methodology of Fair and Sustainable Wellness) that measures the territorial gaps and their predisposition or not to innovation which is an essential prerequisite for the dissemination of AI.

For this purpose, the present contribution unfolds with the following structure:

- description of the theoretical reference framework and of the indicators used concerning economic well-being, training, research and development and business innovation;
- description of the methodology for the construction of the composite indicator to measure the potential regional artificial intelligence;
- description of the results, also through a geo-referenced map of the regions potentially prone to the use of artificial intelligence;
- conclusions.

Keywords Artificial intelligence · Index · Regions

D. Marino (✉)
Mediterranea University of Reggio Calabria, 89100 Reggio Calabria, Italy
e-mail: dmarino@unirc.it

D. Tebala
National Institute of Statistics, 88100 Catanzaro, Italy
e-mail: tebala@istat.it

D. Marino and M. A. Monaca (eds.), *Economic and Policy Implications of Artificial Intelligence*, Studies in Systems, Decision and Control 288,
https://doi.org/10.1007/978-3-030-45340-4_1

1 Introduction

Artificial Intelligence (from now on AI) is synonymous with innovation. In the laboratories of companies and universities, we start to use more innovative technologies to answer specific questions, but in Italy IA is not very widespread. As Gerhard Friedrichn, professor of computer science, dean of the Faculty of Technical Sciences of the University of Klagenfurt, one of the pioneers of research about AI in Europe, says, "artificial intelligence will not come in the coming years, but it will eventually come. There is nothing that can prevent it." Therefore, for the purpose of a better understanding of the phenomenon to date, this study analyzes the potential use of artificial intelligence in the Italian regions through a composite index (using the methodology of Fair and Sustainable Wellness) that measures the territorial gaps and their predisposition or not to innovation which is an essential prerequisite for the dissemination of AI.

For this purpose, the present contribution unfolds with the following structure:

- description of the theoretical reference framework and of the indicators used concerning economic well-being, training, research and development and business innovation;
- description of the methodology for the construction of the composite indicator to measure the potential regional artificial intelligence;
- description of the results, also through a geo-referenced map of the regions potentially prone to the use of artificial intelligence;
- conclusions.

2 Process to Calculate the Indicator

2.1 Description of the Theoretical Framework

The approach used involves the construction of macro areas (pillars) by aggregating elementary indicators. Both pillars and elementary indicators have been considered non-replaceable. To construct synthetic index, we adopted the following indicators and polarity [1] (Table 1):

Table 1 Indicators and polarity

Macro areas	Indicators	Polarity
Economic well-being [2]	GDP per capita Gross fixed investments (ratio between investments and gross domestic product per 100)	+
Training, research and development [2]	Research and development employees (per 1000 inhabitants) Total expenditure for research and development (as a percentage of GDP) Graduates in technical and scientific disciplines (per 1000 residents aged 20–29)	+
Business innovation [2]	Companies that have introduced at least one product or process innovation (percentage of total companies) Companies with cooperation agreements for innovation (percentage of companies with innovative product/ process activities) Companies with innovative activities (percentage of total enterprises) % innovative startups on the total of new joint-stock companies	+

3 Methodology

The matrix relating to data on Italian provinces was divided into four progressive steps:

- Selection of a set of basic indicators on the basis of an ad hoc evaluation model hinging upon the existence of quality requirements;
- Further selection aimed at balancing the set of indicators within the theoretical framework of the structure. Outcome indicators are impact indicators as the ultimate result of an action as a result of a stakeholder activity or process;
- Calculation of synthetic indices (pillars), by making use of the methodology proved more appropriate to obtain usable analytical information;
- Processing of a final synthetic index as a rapid empirical reference concerning the degree of potential Artificial Intelligence of Italian provinces.

Missing values were attributed via the *hot-deck* imputation and, where not possible, with Italy's average value.

The choice of the synthesis method is based on the assumption of a formative measurement model, in which it is believed that the elementary indicators are not replaceable, which is to say, cannot compensate each other.

The exploratory analysis of input data was performed by calculating the mean, average standard deviation and frequency, as well as correlation matrix and principal component analysis. Since this is a non-compensatory approach, the simple aggregation of elementary indicators was carried out using the correct arithmetic average with a penalty proportional to the "horizontal" variability.

Normalization of primary indicators took place by conversion into relative indexes compared to the variation range (min–max).

Attribution of weights to each elementary indicator has followed a subjective approach, opting for the same weight for each of them. Since, in some cases, the elementary indicators showed different polarity, it was necessary to reverse the sign of negative polarities by linear transformation.

For the synthetic indicator calculation, we used the *Adjusted Mazziotta-Pareto Index* (AMPI), which is used for the min–max standardization of elementary indicators and aggregate with the mathematical average penalized by the "horizontal" variability of the indicators themselves. In practice, the compensatory effect of the arithmetic mean (average effect) is corrected by adding a factor to the average (penalty coefficient) which depends on the variability of the normalized values of each unit (called horizontal variability), or by the variability of the indicators compared to the values of reference used for the normalization.

The synthetic index of the i-th unit, which varies between 70 and 130, is obtained by applying, with negative penalty, the correct version of the penalty method for variation coefficient (AMPI $\pm$), where:

$$AMPIi- = Mri - Sricvi \tag{1}$$

where Mri e Sri are, respectively, the arithmetic mean and the standard deviation of the normalized values of the indicators of the i unit, and $cvi = Sri/Mri$ is the coefficient of variation of the normalized values of the indicators of the i unit.

The correction factor is a direct function of the variation coefficient of the normalized values of the indicators for each unit and, having the same arithmetic mean, it is possible to penalize units that have an increased imbalance between the indicators, pushing down the index value (the lower the index value, the lower the level of potential AI).

This method satisfies all requirements for the wellbeing synthesis [1]:

- Spatial and temporal comparison
- Irreplaceability of elementary indicators
- Simplicity and transparency of computation
- Immediate use and interpretation of the obtained results
- Strength of the obtained results.

Table 2 Mean, σ and frequency macro areas

	Economic well-being	Training, research and development	Business innovation
Mean	100.024	94.152	93.019
σ	14.7	13.078	11.834
Frequency	20	20	20

An influence analysis was also performed to assess the robustness of the method and to verify if and with which intensity the composite index rankings change following elimination from the starting set of a primary indicator. This process has also permitted us to analyze the most significant indicators.

The analysis was conducted using the *COMIC* (Composite Indices Creator) software, developed by ISTAT. The software allows calculating synthetic indices and building rankings, as well as easily comparing different synthesis methods to select the most suitable among them, and write an effective report based upon results.

4 Description of the Results

Tables 2 and 3 reveal a moderate variability while Tables 4 and 5 show significant correlations between the analyzed indicators of the macro areas: direct correlation between Business Innovation index and Economic Well-Being index ($r = 0.703$) and between Business Innovation index and Training, Research And Development index ($r = 0.345$). In particularly, there are significant direct correlations between Companies with innovative activities and Companies that have introduced at least one product or process innovation, between Total expenditure for research and development and Research and development employees and between Research and development employees and GDP per capita,

The influence analysis describes the indicators that most influence the composition of rosters in tourism of provinces. In analyzing Tables 6 and 7, we can see that most significant macro area is Economic Well-Being (mean = 1.5, $\sigma = 1.8$) and the most important indicator is Gross fixed investments (mean = 2.5, $\sigma = 2.1$).

5 Discussion and Conclusions

The cartographic representation of the final composite index value, alongside with the descriptive analysis of data, yields the usual dualistic pattern South/Center-North of Italy as in other domains of BES (Table 8 and Fig. 1).

In particular, the best performances are grouped in Friuli-Venezia Giulia, Lombardy and Piedmont, but the region that can potentially use artificial intelligence

Table 3 Mean, σ and frequency indicators

	GDP per capita	Gross fixed investments	Research and development employees	Companies that have introduced at least one product or process innovation	Companies with cooperation agreements for innovation	Companies with innovative activities	Graduates in technical and scientific disciplines	Total expenditure for research and development	% innovative startups on the total of new joint-stock companies
Mean	25.294.925	18.4	4.09	31,425	12,855	44.1	14,095	1.13	3.1
σ	6.623.961	3.269	2.114	6.988	3.28	6.93	2297	0.479	1.069
Frequency	20	20	20	20	20	20	20	20	20

Table 4 Correlation matrix of the macro areas

Macro areas	Economic well-being	Training, research and development	Business innovation
Economic well-being	1.000		
Training, research and development	0.281	1.000	
Business innovation	**0.345**	**0.703**	1.000

Table 5 Correlation matrix of indicators

Indicators	GDP per capita	Gross fixed investments	Research and development employees	Companies that have introduced at least one product or process innovation	Companies with cooperation agreements for innovation	Companies with innovative activities	Graduates in technical and scientific disciplines	Total expenditure for research and development	% innovative startups on the total of new joint-stock companies
GDP per capita	1.000								
Gross fixed investments	0.288	1.000							
Research and development employees	**0.828**	−0.045	1.000						
Companies that have introduced at least one product or process innovation	**0.513**	−0.112	**0.734**	1.000					
Companies with cooperation agreements for innovation	0.095	−0.452	0.144	0.086	1.000				

(continued)

Table 5 (continued)

Indicators	GDP per capita	Gross fixed investments	Research and development employees	Companies that have introduced at least one product or process innovation	Companies with cooperation agreements for innovation	Companies with innovative activities	Graduates in technical and scientific disciplines	Total expenditure for research and development	% innovative startups on the total of new joint-stock companies
Companies with innovative activities	**0.518**	– 0.199	**0.771**	**0.924**	0.019	1.000			
Graduates in technical and scientific disciplines	−0.165	−0.175	0.014	0.342	−0.189	0.324	1.000		
Total expenditure for research and development	**0.559**	−0.206	**0.893**	**0.619**	0.111	**0.734**	0.082	1.000	
% innovative startups on the total of new joint-stock companies	**0.609**	**0.577**	0.304	0.289	−0.056	0.146	−0.094	0.011	1.000

Table 6 Influence analysis: mean and σ of the shifts for basis indicator of macro areas

Macro areas	Mean	σ
Economic well-being	**1.500**	**1.830**
Training, research and development	1.600	1.251
Business innovation	1.500	1.396
Mean	1.533	1.493
σ	0.047	0.246

Table 7 Influence analysis: mean and σ of the shifts for basis indicator of indicators

Indicators	Mean	σ
GDP per capita	1.609	1.748
Gross fixed investments	**2.528**	**2.111**
Research and development employees	0.690	1.025
Companies that have introduced at least one product or process innovation	0.690	1.025
Companies with cooperation agreements for innovation	2.069	1.206
Companies with innovative activities	1.379	1.305
Graduates in technical and scientific disciplines	2.069	1.719
Total expenditure for research and development	0.919	1.305
% innovative startups on the total of new joint-stock companies	1.839	2.194
Mean	1.533	1.493
σ	0.047	0.246

in the most widespread manner is Emilia Romagna (Index 111.23), thanks mainly to research workers and development (8.4 per 1000 residents.) and to companies that have introduced at least one product or process innovation (43.9% of total companies).

Sicily occupies the last position of the ranking (Index 80.9) and follows the negative trend of Southern Italy expect Abruzzo (11th place—Index 93.0) and Basilicata (13th place—Index 89.4) with a good number of graduates (respectively 17.2 and 15.9 per 1000 ab. Aged 20–29 years—average Italy 13.8) and a good ratio between investments and GDP which measures the propensity to invest and that is equal to 22.2 and 23%—average Italy 17.2%).

Our systemic index would be useful to have an idea of the general status of IA and to guide government actions.

Table 8 Synthetic index ranking of potential IA

Regions	Value	Rank
Emilia Romagna	111.235	1
Friuli Venezia Giulia	110.777	2
Lombardy	107.740	3
Piedmont	106.367	4
Veneto	103.425	5
Marche	102.481	6
Lazio	99.175	7
Trentino Alto Adige/Sudtirol	98.786	8
Liguria	97.715	9
Tuscany	96.743	10
Abruzzo	93.014	11
Umbria	92.954	12
Basilicata	89.371	13
Puglia	84.695	14
Valle d'Aosta / Vallée d'Aoste	84.293	15
Campania	84.033	16
Molise	82.133	17
Calabria	81.053	18
Sardinia	80.928	19
Sicily	80.914	20
Italy	*100.000*	

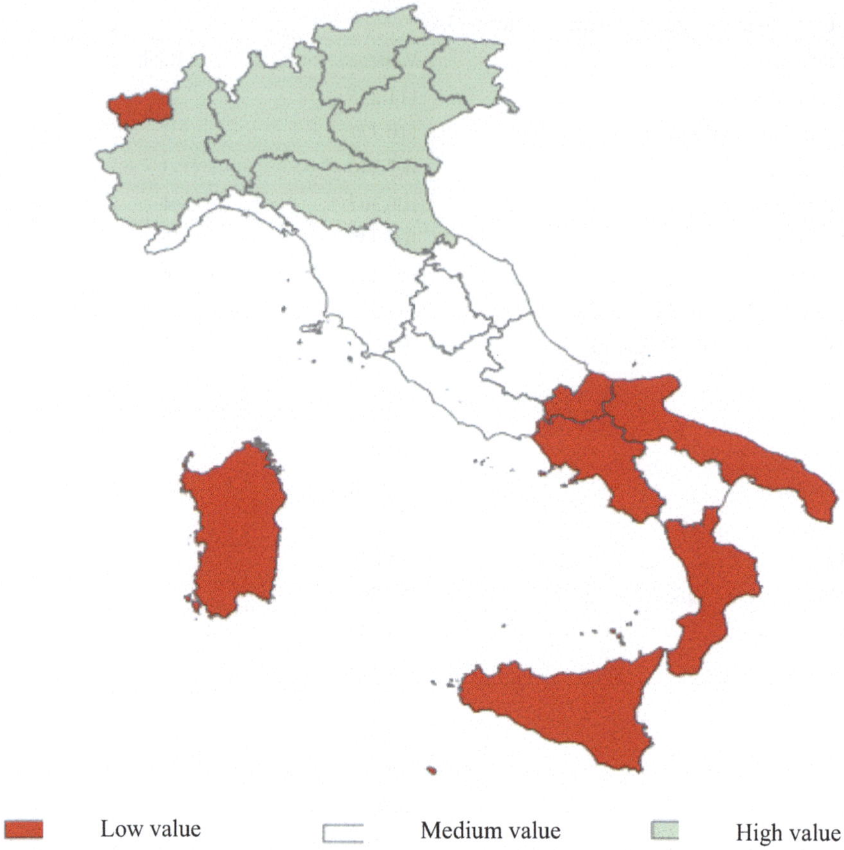

Fig. 1 Synthetic Index ranking of potential IA for regions

References

Istat.: (2018) UrBes 2018—Il benessere equo e sostenibile in Italia https://dati.istat.it

Hybrid Fuzzy Differential System and Artificial Neural Networks: Some Issues in Economics

Massimiliano Ferrara, Iside Rita Laganà, and Domenica Stefania Merenda

Abstract Hybrid systems evolve in continuous time, like differential systems, but undergoing a fundamental change in their governing equations at a sequence of discrete times. When a continuous time dynamics of a hybrid system is given by a Fuzzy Differential Equations (FDEs), the system is called a hybrid fuzzy differential system. This structural and mathematical framework in our opinion is very close to the context of Artificial Intelligence and its applications in various fields.

Keywords Hybrid differential equation · Fuzzy differential equations · Artificial intelligence

Mathematics Subject Classification (2020) 34K34 hybrid systems · 34K36 fuzzy functional differential equations · 94CXX circuits networks

1 Introduction

In the recent years, fuzzy analysis and fuzzy differential equations were proposed to handle uncertainty due to incomplete information that appears in many mathematical or computer models of some deterministic real world phenomena. This theory has been further developed and a wide number of applications of this theory have been considered in the real-world systems. This theory is attractive because it is based on a very intuitive, although somewhat subtle, idea capable of generating many intellectually appealing results that provide new insights to old, often-debated

M. Ferrara (✉)
I—CRIOS, University Bocconi, Milan, Italy
e-mail: Massimiliano.ferrara@unirc.it

M. Ferrara · I. R. Laganà · D. S. Merenda
Department of Law, Economics and Human Sciences and Decisions Lab,
University Mediterranea of Reggio Calabria, Reggio Calabria, Italy
e-mail: isiderita.lagana@unirc.it

D. S. Merenda
e-mail: stefania.merenda@unirc.it

D. Marino and M. A. Monaca (eds.), *Economic and Policy Implications of Artificial Intelligence*, Studies in Systems, Decision and Control 288,
https://doi.org/10.1007/978-3-030-45340-4_2

questions. Thereafter, the applications of this interesting theory are described in the mathematical modeling and show us the way to model the physical problems with fuzzy parameters. In the final step, some important numerical techniques are prescribed to solve fuzzy differential equations. Fuzzy differential equations (FDEs) are an interesting field of the analysis mathematics very useful for studying and solving large proportions of problems in many issue of applied mathematics, e.g. physics, geography, medicine, biology, control chaotic systems, bioinformatics and computational biology, synchronize hyperchaotic systems, economics and finance, and so on. See for instance Casasnovas et al. (2005), Feng and Chen (2005), Zhang et al. (2005). FDEs born in order to model the propagation of epistemic uncertainty in a dynamical environment. In fact FDEs are significant to model problems where the degree of ambiguity is high (Otadi et al. 2017). FDEs can be studied by several approaches. The Hukuhara differentiability for fuzzy number valued functions was the first approach which has been utilized. Fuzzy differential equations were first formulated by Kaleva (1987) and Seikkala (1987) in time dependent form. A very general formulation of a fuzzy first-order initial value problem, has been given by Buckley and Feuring (2000).

Hybrid systems evolve in continuous time, like differential systems, but undergoing a fundamental change in their governing equations at a sequence of discrete times. When a continuous time dynamics of a hybrid system is given by a FDE, the system is called a hybrid fuzzy differential system. For analytical results on hybrid fuzzy differential equations (HFDEs), see Lakshmikantham and Liu (1998) and Lakshmikantham and Mohapatra (2003). Hybrid systems are devoted for modeling, designing, and validating interactive systems of computer programs and continuous systems as well. These control systems which are capable of controlling complex systems have discrete event dynamics as well as continuous time dynamics that can be modeled by hybrid systems.

In this paper, we propose an approach to solve the hybrid fuzzy differential equations based on the feed-forward neural networks. This method shows that using neural networks provides solutions with good generalization and the high accuracy. To do this we shall briefly discuss the following subjects; Artificial neural networks (ANN), and FDEs. (ANN) are computing systems that are inspired by, but not identical to, biological neural networks that constitute animal brains. Such systems "learn" to perform tasks by considering examples, generally without programmed with task-specific rules. PDEs are relevant to approach all the phenomena that need modeling with uncertain parameters. It gives us enough tools to model real-world system and approach us much closer to its behavior.

2 The Hybrid Fuzzy Differential System: A New Model

Consider the following hybrid fuzzy differential system

$$\begin{cases} y'(x) = f(x, y(x), \lambda_k(y_k)), & x \in [x_k, x_{k+1}] \\ y(x_k) = y_k \end{cases} \tag{1}$$

where $0 \leq x_0 < x_1 < \cdots < x_k < \cdots, x_k \to \infty, f \in C[\mathbb{R}^+ \times E \times E, E], \lambda_k \in C[E, E]$

To be specific, the system would look like

$$y'(x) = \begin{cases} y_0'(x) = f(x, y_0(x), \lambda_0(y_0)), \; y_0(x_0) = y_0, & x_0 \leq x \leq x_1 \\ y_1'(x) = f(x, y_1(x), \lambda_1(y_1)), \; y_1(x_1) = y_1, & x_1 \leq x \leq x_2 \\ \vdots & \vdots \\ y_k'(x) = f(x, y_k(x), \lambda_k(y_k)), \; y_k(x_k) = y_k, & x_k \leq x \leq x_{k+1} \\ \vdots & \vdots \end{cases} \tag{2}$$

Assuming that the existence and uniqueness of solutions of Eq. (1) hold for each $[x_{k,}x_{k+1}]$, by the solution of Eq. (1), we obtain the following function

$$y(x) = y(x, x_0, y_0) = \begin{cases} y_0(x), & x_0 \leq x \leq x_1 \\ y_1(x), & x_1 \leq x \leq x_2 \\ \vdots & \vdots \\ y_k(x), & x_k \leq x \leq x_{k+1} \\ \vdots & \vdots \end{cases} \tag{3}$$

We note that the solutions of Eq. (1) are piecewise differentiable in each interval for $x \in [x_k, x_{k+1}]$, for a fixed $y_k \in E$ and $k = 0, 1, 2, \ldots$. We replace Eq. (1) by the following equivalent system

$$\begin{cases} \underline{y}'(x) = \underline{f}(x, y(x), \lambda_k(y_k)), \; \underline{y}(x_k) = \underline{y_k} \\ \overline{y}'(x) = \overline{f}(x, y(x), \lambda_k(y_k)), \; \overline{y}(x_k) = \overline{y_k} \end{cases} \tag{4}$$

The parametric form of the above systems is given by

$$\begin{cases} \underline{y}'(x, r) = F\Big(x, \underline{y}(x, r), \overline{y}(x, r), \underline{\lambda_k}(y_k)(r), \overline{\lambda}_k(y_k)(r)\Big), \; \underline{y}(x_k, r) = \underline{y_k}(r) \\ \overline{y}'(x, r) = G\Big(x, \underline{y}(x, r), \overline{y}(x, r), \underline{\lambda_k}(y_k)(r), \overline{\lambda}_k(y_k)(r)\Big), \; \overline{y}(x_k, r) = \overline{y}_k(r) \end{cases} \tag{7}$$

where $x \in [x_k, x_{k+1}]$ and $r \in [0, 1]$. Using Bede's characterization theorem proposed by Bede (2008), Pederson and Sambandham (2009) generalized the following characterization theorem for HFDEs.

Theorem 1 *Consider the HFDE* (1) *expanded as Eq.* (2) *where for* $k = 0, 1, 2, \ldots$ *and each* $f_k\colon [x_k, x_{k+1}] \times E \to E$, *we have*:

1. $[f_k(x,y)]^r = \left[\underline{f}^r_k\left(x,\underline{y},\overline{y}\right), \overline{f}^r_k\left(x,\underline{y},\overline{y}\right)\right]$;
2. $\underline{f}^r_k$ and $\overline{f}^r_k$ *are equicontinuous and uniformly bounded on any bounded set;*
3. *There exists a* $L_k > 0$ *such that*

$$\left|\underline{f}^r_k(x,y,z) - \underline{f}^r_k(x,y_1,z_1)\right| \le L_k \max\{|y_1 - y|, |z_1 - z|\} \text{ for all } r \in [0,1]$$

$$\left|\overline{f}^r_k(x,y,z) - \overline{f}^r_k(x,y_1,z_1)\right| \le L_k \max\{|y_1 - y|, |z_1 - z|\} \text{ for all } r \in [0,1]$$

Then, the FIVP (Eq. 1) and system of ODEs

$$\begin{cases} \left(\underline{y}^r_k(x)\right)' = \underline{f}^r_k\left(x,\underline{y}^r_k,\overline{y}^r_k\right) \\ \left(\overline{y}^r_k(x)\right)' = \overline{f}^r_k\left(x,\underline{y}^r_k,\overline{y}^r_k\right) \\ \underline{y}^r_k(x_k) = \underline{y}^r_{k-1}(x_k), \quad \text{if } k>0, \quad \underline{y}^r_0(x_0) = \underline{y}^r_0 r \\ \overline{y}^r_k(x_k) = \overline{y}^r_{k-1}(x_k), \quad \text{if } k>0, \quad \overline{y}^r_0(x_0) = \overline{y}^r_0 r \end{cases}$$

are equivalent.

3 Neural Networks: New Issues

Using neural networks provides solutions with very good generalizability (such as differentiability). However, an important future of multilayer perceptrons is their utility to approximate functions, which leads to a wide applicability in most problems. In this article, the function approximation capability of feed-foreword neural networks is used by expressing the trial solution for system (7) as the sum of two terms (see Eq. 6). The first term satisfies the initial conditions and does not contain adjustable parameters. The second term involves a feed-forward neural network to be trained, so satisfies the differential equations. Since it is known as a multilayer perceptron with one hidden layer which can approximate any function to arbitrary accuracy, the multilayer perception is used as the type of the network architecture.

If $\underline{y}_T(x, r, \underline{pcorrespondingnetworksare})$ is a trial solution for the first equation in system (1) and $\overline{y}_T(x,r,\overline{p})$ is a trial solution for the second equation in system (1) where $\underline{p}$ and $\overline{p}$ are adjustable parameters (indeed $\underline{y}_T(x,r,\bar{p})$ and $\overline{y}_T(x,r,\bar{p})$ are approximations of $\underline{y}_T(x,r)$ and $\overline{y}_T(x,r)$ respectively), then a discretized issue of system (1) might be converted to the optimization problem

$$\min_{\vec{v}} \sum_{i=1}^{m} \left\{ \begin{array}{l} \left(\underline{y}'_T(x_i, r, \underline{v}) - F\left[x_i, \underline{y}_T(x_i, r, \underline{v}), \overline{y}_T(x_i, r, \overline{v}), \overline{\lambda}_k(y_k)(r), \overline{\lambda}_k(y_k)(r)\right]\right)^2 \\ +\left(\underline{y}'_T(x_i, r, \overline{v}) - G\left[x_i, \underline{y}_T(x_i, r, \underline{v}), \overline{y}_T(x_i, r, \overline{v}), \underline{\lambda}_k(y_k)(r), \overline{\lambda}_k(y_k)(r)\right]\right)^2 \end{array} \right\} \tag{5}$$

were $\vec{v} = (\underline{v}, \overline{v})$ includes all adjustable parameters with the initial conditions

$$\underline{y_T}(x_0, r, \underline{v}) = \underline{y_0}(r), \quad \overline{y}_T(x_0, r, \overline{v}) = \overline{y}_0(r)$$

Each trial solution $\underline{y}_T$ and $\overline{\mathrm{y}}_{\mathrm{T}}$, employs one feed-forward neural network for which the corresponding networks are denoted by $\underline{N}$ and $\overline{N}$, with adjustable parameters $\underline{v}$ and $\overline{v}$, respectively. Thus, $\underline{y}_T$ and $\overline{y}_T$ can be selected as follows

$$\begin{array}{l} \underline{y}'_T(x, r, \underline{v}) = \underline{y}(x_0, r) + (x - x_0)\underline{N}(x, r, \underline{v}) \\ \overline{y}'_T(x, r, \overline{v}) = \overline{y}(x_0, r) + (x - x_0)\overline{N}(x, r, \overline{v}) \end{array} \tag{6}$$

where $\underline{N}$ and $\overline{N}$ are single-output feed-forward neural networks with adjustable parameters $\underline{v}$ and $\overline{v}$, respectively. Here, x and r are the network inputs. It is easy to see that in Eq. (6), $\underline{y}_T$ and $\overline{y}_T$ satisfy the initial conditions. From Eq. (6), it is easy to show that

$$\begin{cases} \underline{y}'_T(x, r, \underline{v}) = \underline{N}(x, r, \underline{v}) + (x - x_0)\frac{\partial \underline{N}}{\partial x} \\ \underline{y}'_T(x, r, \overline{v}) = \overline{N}(x, r, \overline{v}) + (x - x_0)\frac{\partial \overline{N}}{\partial x} \end{cases} \tag{7}$$

Now suppose a multilayer perceptron has a hidden layer with H sigmoid units and a linear output unit. Therefore, we have

$$\begin{cases} \underline{N} = \sum_{i=1}^{m} \underline{w}_i \sigma(\underline{t}_i), \quad \underline{t}_i = \underline{a}_{i1} x + \underline{a}_{i2} r + \underline{bi} \\ \overline{N} = \sum_{i=1}^{m} \overline{w}_i \sigma(\overline{t}_i), \quad \overline{t}_i = \overline{a}_{i1} x + \overline{\alpha}_{i2} r + \overline{bi} \end{cases} \tag{8}$$

were $\sigma(t)$ is the sigmoid transfer function, $\underline{a}$ and $\overline{a}$ ($m \times 2$ matrices) are the weights of input layers, and $\underline{b}$ and $\overline{b}$ ($m \times 1$ matrices) are the bias vectors of input units $\underline{w}$ and $\overline{w}$ ($m \times 1$ matrices) are the weight vectors of output units, and $\sigma(t) = 1/\left(1 + e^{-t}\right)$ is the sigmoid transfer function. The following is obtained

$$\begin{cases} \frac{\partial \underline{N}}{\partial x} = \sum_{i=1}^{m} \underline{w}_i \underline{\alpha}_{i_1} \sigma'(\underline{t}_i) \\ \frac{\partial \overline{N}}{\partial x} = \sum_{i=1}^{m} \overline{w}_i \overline{\alpha}_{i_1} \sigma'(\overline{t}_i) \end{cases} \tag{9}$$

where $\sigma'(\bar{t}_i)$ is the first derivative of the sigmoid function. Now, if we substitute Eq. (7) in (5), the constrained optimization problem (5) might be changed with the unconstrained optimization problem as follow

$$\min_{\vec{v}} \sum_{i=1}^{n} \left\{ \begin{array}{l} \left(\underline{N}(x, r, \underline{v}) + (x - x_0)\frac{\partial \underline{N}}{\partial x} - F\left[x_i, \underline{y}_T(x_i, r, \underline{v}), \overline{y}_T(x_i, r, \overline{v}), \underline{\lambda}_k(y_k)(r), \overline{\lambda}_k(y_k)(r)\right]\right)^2 \\ +\left(\overline{N}_T(x_{i_1} r, \overline{v}) + (x_i - x_0)\frac{\partial \overline{N}}{\partial x} - G\left[x_i, \underline{y}_T(x_i, r, \underline{v}), \overline{y}_T(x_i, r, \overline{v}), \underline{\lambda}_k(y_k)(r), \overline{\lambda}_k(y_k)(r)\right]\right)^2 \end{array} \right. \tag{10}$$

4 Concluding Remarks and Further Developments in the Frame of Artificial Intelligence and Machine Learning

By this ongoing research we have to show a new method for solving HFDEs. We try introducing the reader to understand the ability of neural networks for approximating the solutions of FDEs. By comparing our achievements with the results obtained using numerical methods, it is clear that our proposed method gives more accurate approximations. Applicability in function approximations of neutral networks is the main reason for using neural networks. More research is in progress for applying and extending this new approach for solving nth-order FDEs as well as a system of FDEs. In this fascinating direction of research we are going to explore the application of this mathematical platform in the frame of Artificial intelligence and Machine Learning. This is the natural extension of the present work.

In summary, we defined a new method for solving HFDEs. We demonstrated the ability of neural networks for approximating the solutions of FDEs. By comparing our achievements with the results obtained using numerical methods, it is clear that our proposed method gives more accurate approximations. Also better results (specially in nonlinear cases) might be possible if we use more neurons or training points. In addition, after solving a FDE, we obtained the solution at any arbitrary point in the training interval (even between training points). Applicability in function approximations of neutral networks is the main reason for using neural networks. More research is in progress for applying and extending this new approach for solving nth-order FDEs as well as a system of FDEs. The numerical results showed that the method has good accuracy and it is efficient summary, we defined a new method for solving HFDEs. We demonstrated the ability of neural networks for approximating the solutions of FDEs. By comparing our achievements with the results obtained using numerical methods, it is clear that our proposed method gives more accurate approximations. Also better results (specially in nonlinear cases) might be possible if we use more neurons or training points. In addition, after solving a FDE, we obtained the solution at any arbitrary point in the training interval (even between training points). Applicability in function approximations of neutral networks is the main reason for using neural networks. More research is in progress for applying and extending this new approach for solving nth-order FDEs as well as a system of FDEs. The numerical results showed that the method has good accuracy and it is efficient. When we talk about Artificial Intelligence, we immediately think of cutting-edge technologies, robots that can understand and decide what actions to take and a futuristic world in which machines and men live together. In reality, Artificial Intelligence, from now AI, and its use are much more real than we can imagine and now used in different areas of daily life.

In technical terms, Artificial Intelligence is a branch of Computer Science that allows the programming and design of both hardware and software systems that allow machines to be equipped with certain characteristics that are typically considered human, such as visual, spatio-temporal and decision-making perceptions. In fact, an intelligent system is created by trying to recreate one or more of these different forms of intelligence which, although often defined as simply human, can actually be traced back to particular behaviors reproducible by some machines.

Starting from the brain work, Artificial Intelligence should be able to perform some human functions, such as:

- act humanly (that is, in an indistinct manner with respect to a human being)
- think humanly (solving a problem with cognitive functions)
- think rationally (that is, using logic as a human being does)
- act rationally (starting a process to obtain the best expected result based on the information available, which is what a human being, often even unconsciously, makes a habit of).

These considerations are very important because they allow us to classify the AI into two great "strands": the weak AI and the strong AI.

The weak formulation claims that a computer will never be able to be equivalent to a human mind, but will only be able to simulate some of the cognitive processes humans without being able to reproduce them in their total complexity. According to the weak setting, the design of smart programs is just a tool to verify theories about how humans could perform cognitive operations. The final purpose of this theory is the construction of machines able to exhibiting behaviors that they would be considered intelligent like humans.

The strong formulation believes that a properly programmed computer can be truly endowed with a pure intelligence, not distinguishable in any way from human intelligence. The idea behind this theory is the concept that goes back to the philosopher English empiricist Hobbes, who argued that "reasoning is nothing else than calculating". The human mind would therefore be the product of a complex set of calculations performed by the brain. According to this conception, a computer properly programmed is a real mind, in the sense that it can be said that the computers in which they were introduced adequate programs understand and have real cognitive states". The debate on the strong formulation of artificial intelligence raises some of the most difficult conceptual problems of all philosophy. It should also be noted that it could very well be to believe in the strong formulation but not to accept the weak formulation. It is perfectly consistent to believe that it is impossible to build machines capable of acting in clever way, but be willing not to recognize such machines in full consciousness if it could be built.

References

Bede, B.: Note on numerical solutions of fuzzy differential equations by predictor corrector method. Inform. Sci. **178**, 1917–1922 (2008)

Buckley, J.J., Feuring, J: Fuzzy differential equations (2000)

Casasnovas, J., Rossello, F.: Averaging fuzzy biopolymers. Fuzzy Set Syst. **152**, 139–158 (2005)

Feng, G., Chen, G.: Adaptive control of discrete-time chaotic systems: a fuzzy control approach. Chaos Soliton Fract. **23**, 459–467 (2005)

Kaleva, O.: Fuzzy differential equations. Fuzzy Sets Syst. **24**, 301–317 (1987)

Lakshmikantham, V., Liu, X.Z.: Impulsive hybrid systems and stability theory. Dyn. Syst. Appl. **7**, 1–10 (1998)

Lakshmikantham, V., Mohapatra, R.N.: Theory of fuzzy differential equations and inclusions. Taylor and Francis, London (2003)

Otadi, et al.: Advances in Fuzzy Logic and Technology (2017)

Pederson, S., Sambandham, M.: Numerical solution of hybrid fuzzy differential equation IVPs by a characterization theorem. Inform. Sci. **179**, 319–328 (2009)

Seikkala, S.: On the fuzzy initial value problem. Fuzzy Sets Syst. **159**, 319–330 (1987)

Zhang, H., Liao, X., Yu, J.: Fuzzy modeling and synchronization of hyperchaotic systems. Chaos Soliton Fract. **26**, 835–843 (2005)

Blockchain and Management Accounting Systems Based on Information Sharing

Domenico Nicolò

Abstract Companies need to share information on the costs and timing of the operations they carry out together with other companies to coordinate and synchronize the activities of each of them. Despite this need, companies avoid sharing information on their operations and activities for fear that they can be modified/changed against them and that competitors can use them to take advantage of them. This paper aims to highlight, from a logical-theoretical point of view, how the inviolability of the blockchain cryptography system can help overcome the resistance of companies to sharing information.

Keywords Blockchain · Management accounting system · Networks

1 Introduction

This short paper is the text of my speech at the AIMED 2019 international conference, held at the Mediterranea University of Reggio Calabria on 27th and 28th September 2019.

As is well-known, blockchain is a shared encrypted database that makes it unmodifiable the data recorded in it. Each data entry is recorded simultaneously in all the nodes of the chain, which has its copy of the database. For this reason, it is impossible to modify the information recorded in it: any attempts to change data in a database would immediately appear as anomalies compared to the other many copies of the database.

This speech does not examine the characteristics of the blockchain and even how it works, moreover widely described by the doctrine and by the technical computer magazines and scientific journals. It aims to highlight a possible useful application of the blockchain to protect the reliability of accounting data entry in the shared accounting systems adopted by companies' part of a network.

D. Nicolò (✉)
Department DIGIES, University Mediterranea of Reggio Calabria, Calabria, Italy
e-mail: domenico.nicolo@unirc.it

D. Marino and M. A. Monaca (eds.), *Economic and Policy Implications of Artificial Intelligence*, Studies in Systems, Decision and Control 288,
https://doi.org/10.1007/978-3-030-45340-4_3

2 The Blockchain as an Opportunity for Successful Innovative Start-Ups

The blockchain and cryptocurrencies play a key role in the business model of many innovative successful start-ups.

The most well-known area of application of the blockchain is cryptocurrencies (Bitcoin, Ethereum, Ripple), but it is also having growing applications in other industries.

Scholars have analyzed the applications of these technologies to various research fields, including:

- big data analysis,
- (IoT) Internet of Things (Zhang and Wen 2015),
- financial services, online payments and reimbursements (Peters et al. 2015; Foroglou and Tsilidou 2015),
- supply and distribution chain,
- risk management,
- copyright management,
- health,
- public services (Akins et al. 2014),
- smart contracts (Kosba et al. 2016),
- reputation building systems (Sharples and Domingue 2016),
- security services (Noyes 2016).

Many innovative startups were created to exploit the business opportunities arising from blockchain and cryptocurrencies in various sectors. Some examples are presented in the following Table 1.

Table 1 Blockchain and start-ups

Slock.its	Connect devices IoT (Internet of Things) to the blockchain with secure interoperability
Gyft	Buy and manage corporate gift cards
Circle	A platform for individuals, institutions and entrepreneurs to build businesses, invest and raise capital with open crypto technologies
Steemit	A platform that rewards the creators of high-quality content to be published in the web
BitTunes	Connects independent artists and buys music using bitcoins
Ubby	A platform that measures the engagement of users' opinions and recommendations and rewards them in Bitcoin when their endorsement to a company allows them to make a sale
Credit dream	Provides a better and cheaper credit option in developing nations
Cryptobuyer CoinPip	Allow users to pay with Bitcoins and companies and receive payment in currency
Isbit	Transfer of funds

3 Outsourcing, Strategic Alliances and Information Sharing Between Companies

In recent years, companies and public administrations tend to outsource operations and activities to increase production quality and efficiency by focusing on the operations they perform most efficiently and at lower prices than other companies (Osborn and Hagedoorn 1997; Garlatti 2000; Ernst and Bamford 2005; Grossi and Mussari 2008). This trend has enhanced collaboration, strategic alliances and transactions between companies (Camagni and Gambarotto 1988; Osland and Yaprak 1995; Soda and Tosi 1998; Lynskey and Yonekura 2001; Lee 2007; Okamuro et al. 2011; Venturelli et al. 2018).

Moreover, to overcome the lack of human, technical and financial resources companies tend to locate their headquarters and operations within industrial districts and clusters (Gilbert et al. 2008). Even within industrial districts and clusters, companies very often collaborate in carrying out operations (Cardoni et al. 2018).

The network that arises from the systematic collaboration between companies provides them with various benefits compared to the big companies, often plagued by bureaucratisation and rigidity. The network has greater agility and adaptability because it is composed of small autonomous companies. Within the network, moreover, small businesses can exchange information, especially in the research field, in this way overcoming another typical weakness of small businesses: the lack of financial resources to be allocated to innovation (Osland and Yaprak 1995; Jagersma 2005).

4 Information System Based on Information Sharing in Interoperating Companies

Companies interoperating need to coordinate and synchronize their activities to carry out the processes and operations they carry out together. Otherwise, their collaboration relationship cannot be successful.

The accounting information system of these companies' part of the network must provide them with information on the costs, the duration, and the start and completion times of these processes carried out jointly with other companies (Mouritsen et al. 2001; Mancini 2011).

If accounting systems are homogeneous and interconnected, transactions between these interoperating companies can take place with a high level of mutual trust, such as internal transfer of goods and services between different responsibility centre of the same company (Hopwood 1996; Osborn and Hagedoorn 1997; Pavan 2019).

This presupposes that the information system of these companies extends beyond the boundaries of its legal structure to reach the strategic boundaries of the network of which they are part. Only in this way the network can be as efficient as large companies.

Network collaboration between companies is generally based on contracts and regulations, in some cases on personal relationships based on trust. At the beginning of the relationship, mutual trust is low and, consequently, the need for information is high. As the parties' exchange information, mutual trust between partner companies grow and so the need for information decreases (Seal et al. 1999; Tomkins 2001; Håkansson and Lind 2004; Bardy 2006; Emsley and Kidon 2007).

Management accounting systems play a fundamental role in creating trust within the network (Seal and Vincent-Jones 1997; Seal et al. 1999; Tomkins 2001; Dekker 2004; Aureli et al. 2019). However, for these tools to work effectively, partners have to share information (Kelley and Thibaut 1978; Mouritsen and Thrane 2006).

Even after creating the relationship of trust between the companies that are part of the network, the accounting systems continue to be important because they provide essential information so that they can collaborate in carrying out processes.

Sharing information on objectives can help network companies, which can exploit synergies with other partner companies, as well as for the solution of common problems (Anderson and Weitz 1992; Heide and Miner 1992; Cannon and Perrault 1999).

The inter-organizational accounting systems based on information sharing play a crucial role also in the internal relations within the network (Lamming 1993; Cannon and Perreault 1999; Dekker 2003; Cooper and Slagmulder 2004; Kajüter and Kulmala 2005; Mouritsen and Thrane 2006) because:

1. induce network companies to implement internal development choices rather than external ones (Mouritsen 1999; Mouritsen and Thrane 2006; Thrane and Hald 2006);
2. they favour the relationships between the companies that are part of the network (Mouritsen et al. 2001; Mouritsen and Thrane 2006; Thrane and Hald 2006) and contribute to the resolution of any conflicts that can arise between them (Chua and Mahama 2007).

5 Future Research Perspectives

A couple of decades ago with the start of strategies for the outsourcing of non-core operations and activities by companies and public administrations, scholars have highlighted the need to adopt management accounting systems based on the sharing of information with partners. Collaboration between companies has become increasingly intense as a result of these outsourcing processes that have made it necessary to control the processes carried out by other companies but interconnected with those carried out internally. However, these systems have not yet been widely used by businesses, probably because they fear that shared information may be altered or may reach competitors who can use it in the competition.

Future research can explore whether the inviolability of blockchain cryptography is allowing the spread of management accounting systems based on information sharing and what are the advantages that these systems offer to the efficiency of the interoperability of companies.

Interesting research evidence can also emerge from experiments with the use of management accounting systems based on blockchain by interoperating companies.

It will also be interesting to see which new business opportunities associated with the use of blockchain and cryptocurrencies will be exploited by innovative startups (Nicolò 2019).

References

Akins, B.W., Chapman, J.L., Gordon, J.M.: A whole new world: income tax considerations of the Bitcoin economy. Pitt. Tax Rev. **12**, 25 (2014)

Anderson, E., Weitz, B.: The use of pledges to build and sustain commitment in distribution channels. J. Mark. Res. **29**(1), 18–34 (1992)

Aureli, S., Cardoni, A., Del Baldo, M., Lombardi, R.: Traditional management accounting tools in SMEs' network. Do they foster partner dialogue and business innovation? Manag. Control. **1**(suppl.), 35–50 (2019)

Bardy, R.: Management control in a business network: new challenges for accounting. Qual. Res. Account. Manag. **3**(2), 161–181 (2006)

Camagni, R., Gambarotto, F.: Gli accordi di cooperazione come nuove forme di sviluppo esterno delle imprese. Economia e politica industriale **58**, 93–138 (1988)

Cannon, J.P., Perreault, W.D., Jr.: Buyer–seller relationships in business markets. J. Mark. Res. **36**(4), 439–460 (1999)

Cardoni, A., Zanin, F., Bartolacci, F., Tompson, G.H.J.: Strategic planning for value creation in business networks. Conceptual framework and theoretical proposals. Manag. Control. **1**, 17–44 (2018)

Chua, W.F., Mahama, H.: The effect of network ties on accounting controls in a supply alliance: field study evidence. Contemp. Account. Res. **24**(1), 47–86 (2007)

Cooper, R., Slagmulder, R.: Interorganizational cost management and relational context. Acc. Organ. Soc. **29**(1), 1–26 (2004)

Dekker, H.C.: Value chain analysis in interfirm relationships: a field study. Manag. Account. Res. **14**(1), 1–23 (2003)

Dekker, H.C.: Control of inter-organizational relationships: evidence on appropriation concerns and coordination requirements. Acc. Organ. Soc. **29**(1), 27–49 (2004)

Emsley, D., Kidon, F.: The relationship between trust and control in international joint ventures: evidence from the airline industry. Contemp. Account. Res. **24**(3), 829–858 (2007)

Ernst, D., Bamford, J.: Your alliances are too stable. Harv. Bus. Rev. **83**(6), 133–141 (2005)

Foroglou, G., Tsilidou, A.L.: Further applications of the blockchain. In: 12th Student Conference on Managerial Science and Technology (2015)

Garlatti, A.: Il riordino delle forme di gestione dei servizi pubblici locali: opportunità e rischi emergenti. Azienda Pubblica **6**, 2000 (2000)

Gilbert, B.A., McDougall, P.P., Audretsch, D.B.: Clusters, knowledge spillovers and new venture performance: an empirical examination. J. Bus. Ventur. **23**(4), 405–422 (2008)

Grossi, G., Mussari, R.: Effects of outsourcing on performance measurement and reporting: the experience of Italian local governments. Public Budgeting Finance **28**(1), 22–38 (2008)

Håkansson, H., Lind, J.: Accounting and network coordination. Acc. Organ. Soc. **29**(1), 51–72 (2004)

Heide, J.B., Miner, A.S.: The shadow of the future: effects of anticipated interaction and frequency of contact on buyer-seller cooperation. Acad. Manag. J. **35**(2), 265–291 (1992)

Hopwood, A.G.: Looking across rather than up and down: on the need to explore the lateral processing of information. Acc. Organ. Soc. **21**(6), 589–590 (1996)
Jagersma, P.K.: Cross-border acquisitions of European multinationals. J. Gen. Manag. **30**(3), 13–34 (2005)
Kajüter, P., Kulmala, H.I.: Open-book accounting in networks. Manage. Acc. Res. **16**(2), 179–204 (2005)
Kelley, H.H., Thibaut, J.W.: Interpersonal relations: a theory of interdependence. Wiley (1978)
Kosba, A., Miller, A., Shi, E., Wen, Z., Papamanthou, C.: Hawk: the blockchain model of cryptography and privacy-preserving smart contracts. In: 2016 IEEE Symposium on Security and Privacy (SP), pp. 839–858. IEEE (2016)
Lamming, R.: Beyond Partnership: Strategies for Innovation and Lean Supply. Hamel Hempstead, Prentice-Hall International (1993)
Lee, C.W.: Strategic Alliances Influence on Small and Medium Firm Performance. J. Bus. Res. **60**(7), 731–741 (2007)
Lynskey, M., Yonekura, S.: Softbank: an internet keiretsu and its leveraging of information asymmetries. Eur. Manag. J. **19**(1), 1–15 (2001)
Mancini, D.: L'azienda-rete e le decisioni di partnership: il ruolo del sistema informativo relazionale. Manag. Control. **1**, 65–97 (2011)
Mouritsen, J.: The flexible firm: strategies for a subcontractor's management control. Acc. Organ. Soc. **24**(1), 31–55 (1999)
Mouritsen, J., Thrane, S.: Accounting, network complementarities and the development of inter-organisational relations. Acc. Organ. Soc. **31**(3), 241–275 (2006)
Mouritsen, J., Hansen, A., Hansen, C.Ø.: Inter-organizational controls and organizational competencies: episodes around target cost management/functional analysis and open book accounting. Manag. Account. Res. **12**(2), 221–244 (2001)
Nicolò, D.: La vulnerabilità delle imprese nella fase di startup. Analisi ed interpretazione delle cause. Giappichelli (2019)
Noyes, C.: Bitav: fast anti-malware by distributed blockchain consensus and feedforward scanning (2016). ArXiv preprint arXiv: 1601.01405
Okamuro, H., Kato, M., Honjo, Y.: Determinants of R&D cooperation in Japanese start-ups. Res. Policy **40**(5), 728–738 (2011)
Osborn, R.N., Hagedoorn, J.: The institutionalization and evolutionary dynamics of interorganizational alliances and networks. Acad. Manag. J. **40**(2), 261–278 (1997)
Osland, G.E., Yaprak, A.: Learning through strategic alliances: processes and factors that enhance marketing effectiveness. Eur. J. Mark. **29**(3), 52–66 (1995)
Pavan, A.: Controllo interno e di gestione nella prospettiva del valore. Manag. Control. **1**, 5–12 (2019)
Peters, G., Panayi, E., Chapelle, A.: Trends in cryptocurrencies and blockchain technologies: a monetary theory and regulation perspective. J. Financ. Perspect. 3(3) (2015)
Seal, W., Vincent-Jones, P.: Accounting and trust in the enabling of long-term relations. Account., Audit. & Account. J. **10**(3), 406–431 (1997)
Seal, W., Cullen, J., Dunlop, A., Berry, T., Ahmed, M.: Enacting a European supply chain: a case study on the role of management accounting. Manag. Account. Res. **10**(3), 303–322 (1999)
Sharples, M., Domingue, J.: The blockchain and kudos: a distributed system for educational record, reputation and reward. In: European Conference on Technology Enhanced Learning, pp. 490–496. Springer, Cham (2016)
Soda, G., Tosi, H.L.: Reti tra imprese: modelli e prospettive per una teoria del coordinamento tra imprese. Carocci (1998)
Thrane, S., Hald, K.S.: The emergence of boundaries and accounting in supply fields: the dynamics of integration and fragmentation. Manag. Account. Res. **17**(3), 288–314 (2006)
Tomkins, C.: Interdependencies, trust and information in relationships, alliances and networks. Acc. Organ. Soc. **26**(2), 161–191 (2001)

Venturelli, A., Caputo, F., Pizzi, S.: L'impatto del contratto di rete nei processi di internazionalizzazione: alcune evidenze empiriche sulle PMI italiane. Manag. Control (2018)
Zhang, Y., Wen, J.: An IoT electric business model based on the protocol of bitcoin. In: 2015 18th International Conference on Intelligence in Next Generation Networks, pp. 184–191. IEEE (2015)

Is the Rapid Increase in Maritime Transports a Threat or an Opportunity for Port Cities? A System Dynamics Analysis

Fabio Carlucci, Andrea Cirà, and Francesca Costanza

Abstract In the last decades the dramatic increase of traffics within the Mediterranean seaports has stimulated the debate on how to design effective infrastructural policies, able to boost ports' efficiency and by this way promoting sustainable development in the surrounding areas. In this regard, system dynamics modelling appears a suitable methodological approach to deal with the management of seaports, characterized by high level of complexity, due to the number of interacting variables, actors and challenges. This paper, focusing on a middle-sized seaport within the Mediterranean basin, proposes a simulation model to portray the causal tissue of seaports' handling capacity, financial performance and economic impact. The system-wide analyses, including stock-and-flow and causal loop diagramming, in pair with simulations of alternative policy measures, reveal the potential of such methodology in supporting local governments' decision making and implementation.

Keywords Sustainability · Seaport management · Policy making · System dynamics

1 Introduction

Until the 1990s, competitiveness and productivity of the infrastructures were very relevant in terms of economic efficiency of maritime transport. Globalization and the diffusion of global supply networks require increasing efficient of logistic systems. Logistic activities are essential for the economic growth and seaports play

F. Carlucci
University of Salerno, Fisciano (SA), Italy
e-mail: fcarlucci@unisa.it

A. Cirà
University of Messina, Messina, Italy
e-mail: acira@unime.it

F. Costanza (✉)
University LUMSA, Rome, Italy
e-mail: f.costanza1@lumsa.it

D. Marino and M. A. Monaca (eds.), *Economic and Policy Implications of Artificial Intelligence*, Studies in Systems, Decision and Control 288,
https://doi.org/10.1007/978-3-030-45340-4_4

a key role for logistic challenge. Modern logistics model requires integrated port networks, in order to match the needs of the new industrial organization. The evolution of the transport sector have deeply changed structure and functions of the main maritime terminals which have now become focal points of logistics chains, so that it is now agreed that ports are key element of global supply networks. Obviously, port management is the prerogative for the strategic development of a port, stimulating dialogue between stakeholders and other ports in order to realize an efficient network. Particularly after containerization revolution, seaports have developed a cluster structure model with several, different decision-making units cooperating in order to supply integrated port services. Cooperation among ports are aimed to plane economic functions achieving scope and scale economies, whereas partnership leads to improvement of promotion and marketing functions. Global supply networks add value to products through just in time supplies (JIT). In this context, seaports are deemed to play a major role in generating logistics efficiency and quality standards of modern logistics systems. Seaports infrastructures are generally financed through public funds. For this reason, appropriate evaluations of direct and indirect impacts of port activities are of great importance in order to support public policy makers.

Investments in seaport infrastructures have several economic effects reducing transport costs, enlarging the marketplace and improving production channels, even if the experiences recorded by the major port networks show that large hub ports serving container have only a marginal economic impact on local systems. Therefore, the analysis of the main examples of port networks and their patterns of operation should be very useful.

The seaport network port can be distinguished on the basis of three types of interdependence: sequential, reciprocal and group relationship. The first kind of relationship between ports is the chain type, based on sequential interdependence in which the output of the first port is the input of the second one and so on. Each port of the chain has a different specialization and produces services for the port that is positioned further down the chain. The ports are linked each other in a value-adding production chain. The second category consists of a set of ports whose relationship is based on mutual interdependence between different nodes. They contribute mutually in an undifferentiated way to the exchange of input and output and together they are able to satisfy customers' needs increasing the overall competitiveness of the group.

The third kind of relationship is based on cooperative competition. The ports that may constitute the network are direct competitors since they share the same market. The relationship that would arise between these ports would be in a cooperative form aimed at enhancing the overall size of the market, creating at the same time, the conditions for mutual growth of their competitiveness compared to other ports. Relationships of mutual complementarity could be created in order to improve the competitiveness, for example, through territorial marketing activities in the field of research for the development of more efficient equipment, and so on.

A port may be positioned indifferently in any one of three patterns described above. Of course, by increasing the size of the network, the range of services will increase, improving the competitive ability and the activities performed by each component of the network.

In particular, in chain type network, the output of a port represents the input of another port. The relationship is more articulated in the case of complex network, in which two ports mutually exchange input and output. Moreover, if there is a port that competes directly with the ports in a network, it needs to have a cooperative attitude, otherwise it would be excluded from the market, because it is less competitive than ports belonging to the network.

Consequently, organizational improvements take on equal, if not greater, importance than infrastructural interventions.

The study focuses the performance management of a middle-sized Mediterranean seaport by adopting a system dynamics perspective (Forrester 1961; Sterman 2000), considered suitable to address the interactions between territorial economy and port's infrastructures and financial performances. So far, two main categories of port-related simulation models have been built in system dynamic literature: a first stream focused the seaport's handling processes (Cheng et al. 2010; Dundovic et al. 2009; Dvornik et al. 2006); a second one faced the relationships between the port's economy and the land's transportation system (Yu et al. 2014; Caballini et al. 2012). Thus, it was evaluated the usefulness of such methodological approach also to pursue the aim of the present research.

2 Materials and Methods

The management of seaports was addressed by adopting a system dynamics perspective, judged adequate to shed light on the complexity of the context under investigation. System dynamics is a simulation methodology, originally developed by Jay Forrester at MIT Institute in Boston (Forrester 1961), and progressively enriched and enlarged as far as applications and modelling approaches (see, among the others, Morecroft 1998; Coyle 2000; Homer and Hirsch 2006; Xing and Dangerfield 2018). SD is suitable for policy analysis and design focusing complex systems, qualified by interdependence, mutual interaction, information feedback, temporal delays, circular causality (Richardson 1991). Through this methodology, a dynamic problem (aggregated pattern occurring over the time and creating concern to some extent) that arises in a social, managerial, economic, or ecological system, can be represented by two kind of diagrams mapping key-variables and their links: stock-and-flow and causal loop diagrams. A stock-and-flow structure is a combination of stocks, flows and information feedbacks, tracking processes of accumulations (Sterman 2000). Three building blocks are present in such diagrams:

(1) Stocks: levels of accumulation of resources (physical, financial, personal, informative) within the reference system. They are graphically represented by mean of rectangles.
(2) Flows: rates, the only items able to directly influence the level of the stocks. Their graphical representation is though pipes feeding (inflows) or draining (outflows) stocks.

(3) Auxiliary: variables aiding calculations (converters) or representing exogenous parameters or constants.

Mathematically a stock-and-flow structure can be formalized as a set of differential equations, which can be inserted in a simulation software to simulate key-variables' behaviors under different conditions. Time is partitioned into discrete intervals of length dt, and simulations allow the system to step over the time one dt at a time. Each stock's level at a certain time, x(t), is the result of its previous value (i.e. calculated at time t-dt) and its net rate of change x'

$$x(t) = x(t - dt) + dt * x'(t - dt) \quad (1)$$

where the net rate of change is the difference between the inflows and the outflows respectively feeding and draining the stock in the chosen time step (Richardson 1991).

Figure 1 shows a simple stock-and-flow structure. Stock 1 has an inflow and an outflow, respectively increasing and reducing its level; stock 2 is linked to a bi-flow, a simplified notation directly calculating the net flow as the difference between in- and out-flows. Together with constants and exogenous variables, each stock is able to influence the flows connected to the other stocks, directly (like stock 1 does with the bi-flow) or through auxiliary variables (e.g. stock 2 affects auxiliary variable 3 and by this way the outflow from stock 1).

An SD model can also be identifiable with a feedback structure, i.e. a network of loops which are circular relationships between variables. These closed causal chains reflect an endogenous point of view (Richardson 2011), in pace with which the internal structure of the model causes the dynamic problem to arise. A useful tool to represent a system's feedback structure is the causal loop diagram. An example of

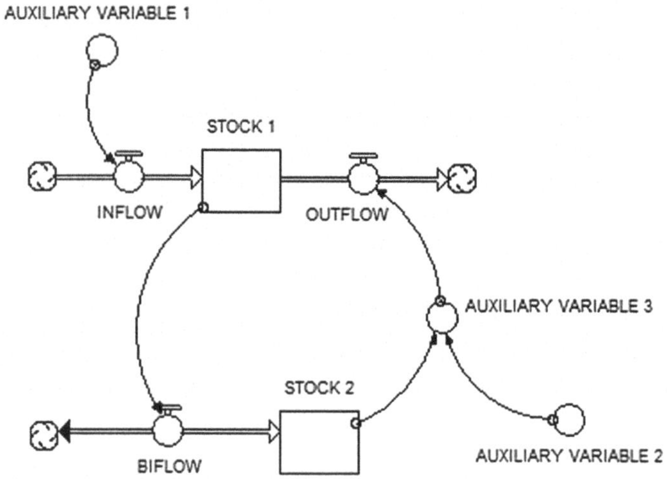

Fig. 1 Example of stock-and-flow diagram

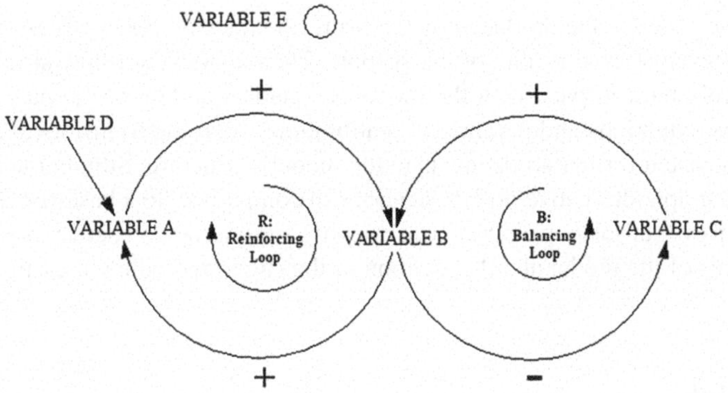

Fig. 2 Example of causal loop diagram

it can be seen in Fig. 2, showing the interaction of two different loops. In this regard, variables A, B and C are endogenous to the model, since they take part to at least one loop, variable D is exogenous, i.e. not included in any loop and can represent either a policy parameter or a constant. On the contrary, auxiliary E is excluded from the model boundaries. Arrows connecting the variables are causal links, denoted by " + " or "−". The sign " + " connotes a positive link: if the independent variable (the cause) increases, *ceteris paribus* the dependent variable (the effect) increases as well; on the other way around, if the cause diminishes so does the effect. The sign "−" is the notation for a negative relation, i.e. that a raise in the independent variable this time determines a reduction in the dependent one (Sterman 2000). Thus, the terms positive and negative are not to be considered as judgments on favorability, but they just indicate if changes in the variables under study go to the same direction (positive links) or to the opposite one (negative links).

In a system dynamics model, the key patterns are the result of interactions between two different kinds of feedback loops:

1. Positive, or reinforcing: these circuits have the tendency to exacerbate initial conditions, provoking virtuous or vicious mechanisms of growth or decay (example, loop R, on the left side of Fig. 2);
2. Negative, or balancing, or counteracting: having the tendency to equilibrate or stabilize the initial dynamics (example, loop B, on the right side of Fig. 2).

Again, the adjectives positive and negative are not to be intended as qualifying wishful or un-wishful dynamics, but just denote a loop's polarity, resulting from the multiplication of all the links' signs. As a consequence, the nature of a loop can be understood by checking the number of negative links: if they are odd, the loop is negative (or balancing, or counteracting), otherwise positive (or reinforcing).

The modelling activities have been divided into main steps. First of all, there was the identification of key-variables and their causal relationships. In particular, for the development of the seaport the following variables were selected: the number of containers that the seaport can handle; the number of firms operating around the seaport; the added value produced by the firms; the demand for goods and services; the costs, revenues and profits of the seaport. Then a stock-and-flow structure was progressively built in pace with the modellers' understanding of the phenomenon. Afterwards, equations and parameters' calibration according to historical data iteratively suggested further adjustments to the model's structure. Simulation activities allowed to graph alternative policy scenarios in comparison to a base-run. Finally, a comprehensive causal loop diagram was drawn to summarize and better communicate key-features of the model and the location of the suggested policy measure.

3 Results

The system dynamics model targets a middle-sized seaport within the Mediterranean Sea a focuses the container business, chosen for its relevance within the European and Mediterranean contexts. The model's structure is made up of five stocks; each of them is fed and/or drained by inflows and outflows or by a bi-flows, creating a feedback structure that portrays the multidirectional exchanges of material, money and information flowing within the port, from the port to the neighbour firms and vice versa. An overview of the model structure is proposed in Fig. 3.

The system has both an economic and financial connotation. Regarding the economic aspect, on the left side it is depicted the effect of firms' activities on the served area: a raise in the stock of running 'firms' causes the added value to increase and by this way stimulates the 'demand of good and services'. The latter has a boosting effect of the number of 'firms', which express a demand for containers. In Fig. 4 it is possible to notice a zoom of a portion of the SD model, where the stock of 'potential container' (within the port) is indicated by the firms, so that its inflow 'indicated handling rate' results from the number of firms multiplied by the average 'annual handling per firm'. However, the number of potential containers deriving

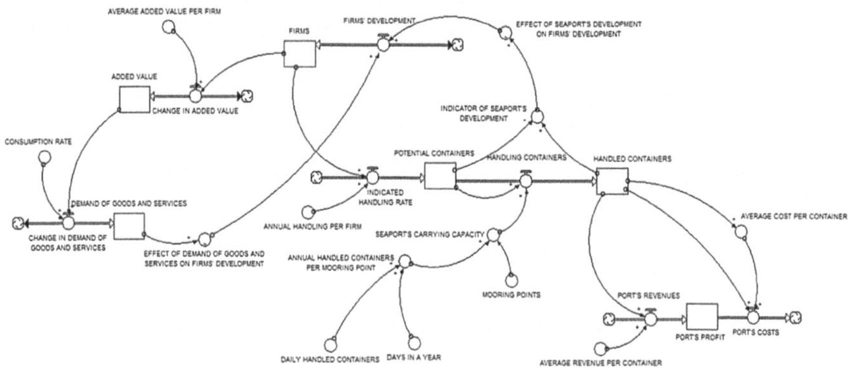

Fig. 3 Overall stock-and-flow diagram of the seaport's model

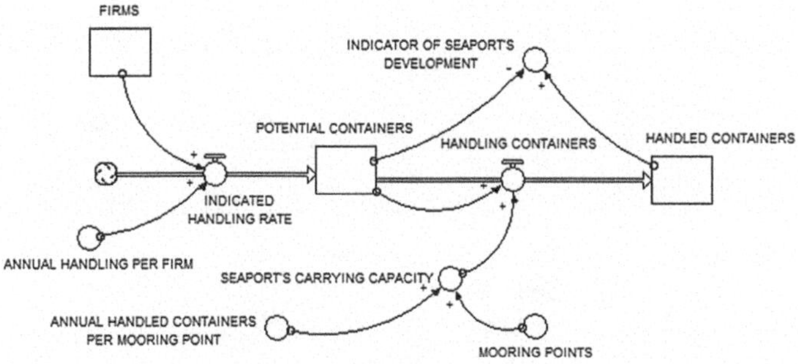

Fig. 4 Seaport's model: zoom of potential containers indicated by firms and constrained by capacity

from firms' demand cannot be fully handled because of the port's limited infrastructures, so that only one part of 'potential containers' passes to the stock of effectively 'handled containers'. This happens through the flow 'handling containers', regulated by the 'seaport carrying capacity' according to the Eq. (2), which takes the minimum between 'potential containers' and 'seaport's carrying capacity':

$$\begin{aligned}&\text{IF (Potential Containers} > \text{Seaport's Carrying Capacity) THEN}\\&\text{(Seaport's Carrying Capacity)ELSE (Potential Containers).}\end{aligned} \tag{2}$$

The model hypothesizes as a base-run condition that the port uses five mooring points, and that each of them can handle 21.900 containers on an annual basis (expressed by the variable 'annual handled container per mooring point', easily calculated by assuming a daily average of 60 handled containers per mooring point). Then the seaport is able to handle about 109.500 containers per year.

On the other side the ratio between the number of seaport's containers and the potential ones, named 'indicator of seaport's development', can be considered as a gauge of the surrounding territory's progressive development: its growth affects the number of firms running in the territory. This is because firms have convenience to locate their plants in the neighbourhood of the seaport in order to reduce transport costs.

In Fig. 5 it is zoomed the financial part of the SD model, where the 'handled containers' contribute to the 'port's revenues' and 'costs', modelled as flows respectively feeding and draining the stock of 'port's profits'.

As obvious, the increase in the containers' traffic generates both higher revenues and costs: each flow is obtained by multiplying the 'handled containers' for the average amounts. So far the 'average revenue per container' is represented by mean of a constant variable, whilst the 'average cost per container' depends upon the number of 'handled containers'. Indeed, it is possible to estimate per each container 110.00 euro of revenues and 80.00 euro of costs, with a step of the average cost per container whenever the threshold of 50,000 handled units is exceeded.

Simulating the base-line scenario of the seaport's cumulated traffic of containers for a 5-years period (Fig. 6), it is possible to realize that in a first period, lasting about

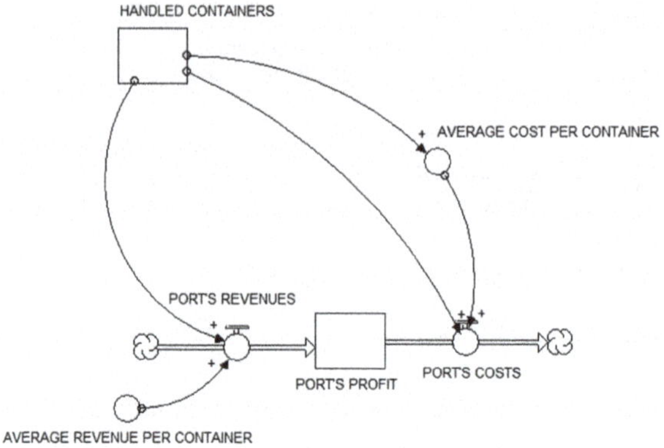

Fig. 5 Seaport's model: zoom of containers traffic's contribution to the financial performance

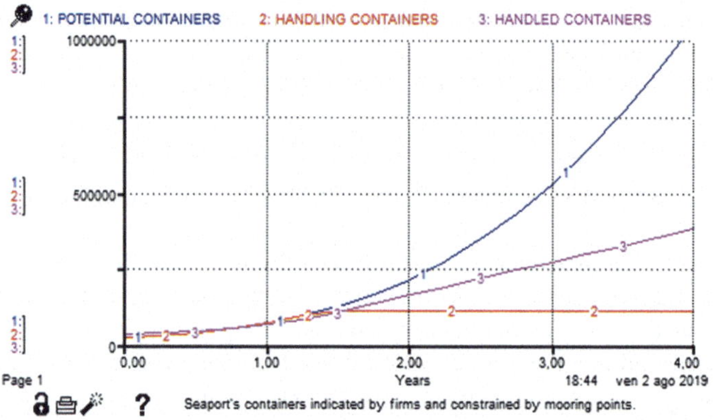

Fig. 6 Simulation of containers indicated by firms and constrained by mooring points

1.5 years, the stocks of 'potential containers' and 'handled containers', as well as the flow 'handling containers' tend to growth exponentially and to overlap with small discrepancies.

This is because the model depicts the mutual and progressive development of the seaport and firms' sub-systems. Indeed, at the beginning of the time frame the firms' development expresses a demand for containers that is growing, but still lower than the seaport's capacity.

When the surrounding businesses' growth is that to determine a relevant demand for containers, the seaport's carrying capacity, constrained by its mooring points, get saturated: the flow of 'handling containers' stands on 109.500 containers (as calculated above) on an annual basis, while the cumulated 'handled containers' shows a steady growth, with an increasing gap with the cumulated 'potential containers'.

In order to better accomplish the demanded traffic of containers, it is possible to hypothesize that the Port Authority decides to invest a portion of its profits to increase the number of handled containers within the seaport. Such policy measure means to increase the number of mooring points. Figure 7 exhibits the impact of alternative investment policies on port's profits, comparing the base-run 5 mooring points (line 1), with scenarios of increased mooring points by three units (line 2) and five units (line 3). In any case each policy, although activated at time 0, starts to become effective after 2.3 years.

It is possible to synthesize the core dynamics of the model through a comprehensive causal loop diagram (Fig. 8) which does not distinguish stocks, flows and auxiliaries, and isolates just 10 variables from the ones belonging to the stock-and-flow structure.

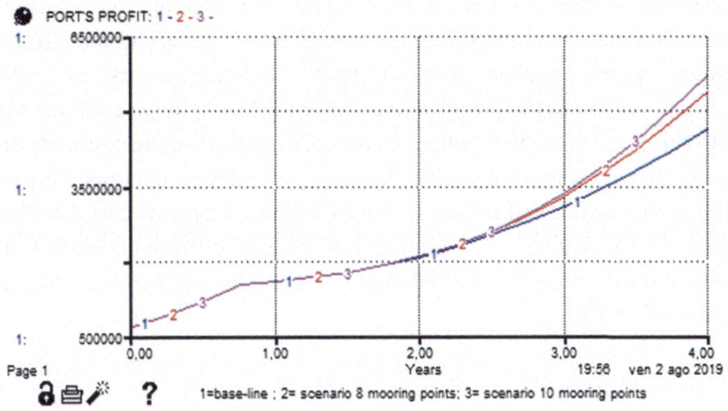

Fig. 7 Port's profits scenarios according with alternative investment policies

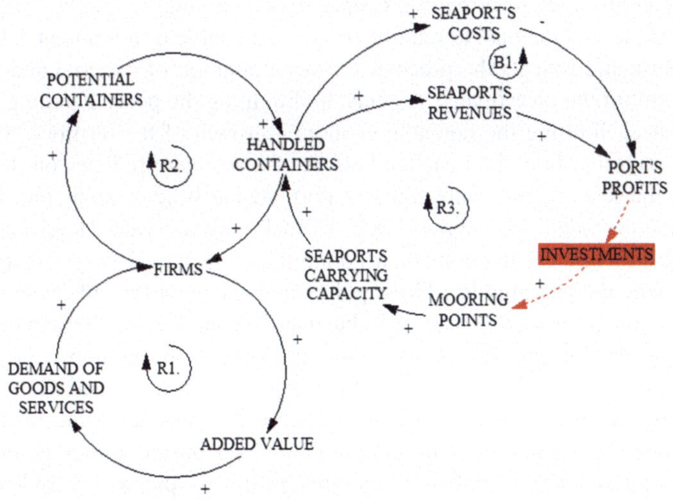

Fig. 8 Overall causal loop diagram of the seaport's model

The diagram is constituted by four loops: three reinforcing (R1, R2 and R3) and one balancing (B1). Loop R1 depicts the self-reinforcing economic dynamics of firms' development through increase of added value and gauge of demand of good and services. On the upper level, loop R2 represents the seaport's traffic, summarized by the variable 'handled containers', driven by two forces, the demand for containers called by firms and indicated by 'potential containers', and the 'seaport's carrying capacity', which in turn is determined by the number of mooring points. The circuits B1 and R3 belong to the financial section of the model. They can be defined "policy loops", since they only exist if an infrastructural investment policy is activated by the Port Authority. For this reason, the variable "investments", not present in the explanatory stock-and-flow structure described above, is highlighted in colour and the new links closing the loops are dashed. While R3 presents the positive effect of an investment on new mooring points, B1 tend to limit such dynamics. Indeed, in R3 more revenues, *ceteris paribus*, make the profits increase; a raise of profits allows more investments on mooring points and enlarge port's capacity, causing the handled containers to raise and thereby revenues to increase again. On the contrary, in B1 more costs have a depleting effect on profits, less profits causes a decrease of investments on mooring points and by this way tend to reduce carrying capacity because of inefficiencies due to a lack of maintenance activities. A diminished carrying capacity is able to reduce the containers handled in the port and by this way the costs associated to the containers' traffic.

4 Discussion and Conclusions

This study contributes to provide a deeper understanding of seaports' infrastructural and financial dynamics in relation to the sustainable development. The stock-and-flow structure reflects the processes of accumulation of material and financial resources within the investigated systems, highlighting the port's handling capacity as a bottleneck limiting the potential economic growth of the territory. Simulated behaviour patterns show the beneficial effect of investment policies on the performances of the seaport, and confirm that 'a port site has little meaning unless capital investment is provided' (Notteboom 2009, p. 24). Finally, the causal loop diagram has the merit to allow a clear and immediate communication of the key system dynamics, thereby having the potential to address policy makers' agendas and decisions.

So far, the outcome model focused the business of containers within a middle-sized seaport of the Mediterranean Sea, addressing infrastructural, economic and financial aspects. Next versions of the model could enlarge the relevant system's boundaries by including other dimensions such as the accessibility to the hinterland and the logistic chain. New markets, other than the container business, may be portrayed, such as the liquid or the dry bulk ones. Future research developments could also face seaports' performance management from an international perspective, for example, considering the European port system as a whole, depicting the routing of goods from port to port, from port to hinterland and vice versa. In this regard, it could be either possible to represent the links between relevant ports within the network, or to consider gate-way regions (grouping several seaports) as unit of analysis.

The results of the analyses highlight market failures and the absence of a policy coordination in the shipping sector. Therefore, it would be desirable a greater policy effort in the use of scientific tools like SD, aimed at the sustainability of seaports and maritime activities.

References

Caballini, C., Sacone, S., Sini, S.: The port as s system of systems: a system dynamics simulation approach. In: Proceedings of the 7th Conference of System of Systems Engineering, Genova, Italy, pp. 191–196, 16–19 July 2012.

Cheng, J. K., Tahar, R. M., Ang, C.-L.: Understanding the complexity of container terminal operation through the development of system dynamics model. Int. J. Shipping Transp. Logistics **2**(4), 429–443 (2010).

Coyle, G.: Qualitative and quantitative modelling in system dynamics: some research questions. Syst. Dyn. Rev. **16**(3), 225–244 (2000). https://doi.org/10.1002/1099-1727(200023)16:3%3C225::AID-SDR195%3E3.0.CO;2-D

Dundovic, C., Bilic, M., Dvornik, J.: contribution to the development of a simulation model for a seaport in specific operating conditions. Promet Traffic Transp. 21(5), 331–340 (2009).

Dvornik, J., Munitic, A., Bilic, M.: Simulation modelling and heuristics optimization of material flow of the port cargo system. Promet Traffic Transp **18**(2), 123–135 (2006)

Forrester, J.W.: Industrial Dynamics. The MIT Press, Cambridge, MA (1961)

Homer, J.B., Hirsch, G.B.: System dynamics modelling for public health: background and opportunities. Am. J. Public Health **96**(3), 452–458 (2006)

Morecroft, J.D.W.: System dynamics and microworlds for policymakers. Eur. J. Oper. Res. **35**(3), 301–320 (1988). https://doi.org/10.1016/0377-2217(88)90221-4

Notteboom, T.: Economic analysis of the European seaport system. Report commissioned by the European Sea Ports Organization (ESPO), ITMMA—University of Antwerp, Belgium (2009).

Richardson, G. P.: System dynamics: simulation for policy analysis from a feedback perspective. In: Fishwick, P.A., Luker, P.A. (eds.) Qualitative Simulation Modeling and Analysis, pp. 144–169. Springer, New York, NY. 10.1007/978–1–4613–9072–5_7 (1991)

Richardson, G.P.: Reflection on the foundations of system dynamics. Syst. Dyn. Rev. **27**(3), 219–243 (2011). https://doi.org/10.1002/sdr.462

Sterman, J.D.: Business Dynamics, System thinking and modeling for a complex world. Mc Graw-Hill, NewYork, NY (2000)

Xing, Y., Dangerfield, B.: Modelling the sustainability of mass tourism in island tourist economies. In: Kunc, M. (eds.) System Dynamics Or Essentials. Palgrave Macmillan, London (2018).

Yu, B., Zhang, C., Kong, L., Bao, H., Wang, W., Ke, S., Ning, G.: System dynamics modeling for the land transportation system in a port city. Simulation **90**(6), 706–716 (2014)

Empowering Individuals, Companies and Organizations to Build Trust Through Rated and Digitalized Reputation

Vincenzo Mancuso, Domenico Marino, Giovanni Mottura and Michele Petrone

Abstract Mevaluate Holding Ltd defines a completely new approach to the reputation. A proprietary algorithm rank reputation through a rating code in the form of A-A-A-100-100 (the topmost score). Each letter/number of this rating pertains to a specific area: criminal, fiscal, civil, studies and training (only for individuals), work and social activity. The rating comes out from weighted facts proven by documents. Whoever wants to have a «digitalized reputational rating» is required to share all the facts with the documented evidence within a user's community. It is not allowed to lie, it is not allowed to exclude information, on pain of registration in "gray list" or "black list". AI plays a big role in three main ways: to check the document veracity and trustworthiness, to define a fine-grained evaluation of the facts illustrated in the text of the document, to anticipate the changes in of individuals and organizations being behavior and refine the weights attributed by the algorithm to the documented facts. An application case (in the economic and judicial field) of AI related to the score expressed by the «digitalized reputational rating» in Italy concerns the so-called «Digital Antimafia» developed in the context of the public private partnership (case study) between:

- National agency for the administration and destination of seized and confiscated assets from organized crime (Anbsc), supervised by the Ministry of the Interior (Italian Government),
- Professional association with traceable reputation auditors (Apart), founded by Mevaluate Holding Ltd, supervised by the Ministry of Economic Development (Italian Government) pursuant to Law 4/2013,

---

V. Mancuso · M. Petrone
Mevaluate Holding Ltd., Dublin, Ireland

D. Marino (✉)
Economic Policy, Mediterranean University of Reggio Calabria, Reggio Calabria, Italy
e-mail: dmarino@unirc.it

G. Mottura
Inag—National Institute for Judicial Administrators, Rome, Italy

M. Petrone
Crop News Onlus, Rome, Italy

D. Marino and M. A. Monaca (eds.), *Economic and Policy Implications of Artificial Intelligence*, Studies in Systems, Decision and Control 288,
https://doi.org/10.1007/978-3-030-45340-4_5

- Crop News (personalized objective reputational reports), online magazine published by Crop News Onlus (founded by Mevaluate Holding Ltd),
- Consumers and Users Federation (Codacons, Codici, Konsumer Italia) and
- National judicial administrators institute (Inag).

Anbsc is searching for new ways to certify the legality status and efficient economic management of seized and confiscated assets even concerning all their economic relationship (suppliers, customers, employees, etc.), guaranteeing the protection of the interests of the consumers and the users.

Keywords Digital reputational rating · Artificial Intelligence

1 Digital Reputational Rating: A New Measure

Mevaluate Holding has defined a completely new approach to the reputation.

A proprietary algorithm ranks the reputation through a rating code—expressed in the form of A-A-A-100-100 (which is the topmost score)—and in Italy it is brought to the attention of the users community in the online magazine Crop News (personalized objective reputational reporting), published by the Crop News Onlus association.

The «digitalized reputational rating» is computed weighting the facts proven by documents. Whoever wants to have a code is required to share all the facts with the documented evidence hosted on the digital infrastructure of reputational qualification documented and traceable Crop News. It is not allowed to lie, it is not allowed to exclude information, on pain of registration in "gray list" or "black list".

Through this digital infrastructure, you can access the «digitalized reputational rating» and decide it's enough for your evaluation but if you need more you can access all the documents (typically for professional usage).

2 How Artificial Intelligence (AI) Improves Rating

To guarantee this result, a validation process about the veracity and trustworthiness of documents is in place as today there is still no professional alternative to a human process. In this scenario AI changes many things.

To check the veracity and trustworthiness of documents is too obvious to be considered. AI must be mostly addressed to extend the measure of the value we assign to a fact through semantic analysis or to foresee cultural changes proposing new ways to interpret and weight facts through time in terms of developing the «digitalized reputational rating».

2.1 Evaluation of Facts

While criminal, fiscal, or civil facts (the first three letters of the «digitalized reputational rating», i.e. A-A-A-) may be sufficiently measured by the consequences (usually negative ones) in conviction years or fines or whatsoever, the same cannot be applied for «studies and training» or «work and social activity».

Just for example (the second number of the «digitalized reputational rating», in example 100), let us consider the score for «work and social activity» area: how we could differentiate two references?

We should delegate someone to «translate» what is written in each letter into, let's say, a five-star evaluation about the judgment the writer of the letter is giving.

But we cannot pretend to have the same person to evaluate every letter for every country (or language) taking into consideration also the natural cultural differences and the reputational rating needs not to be influenced by personal feelings.

AI automatize this process when applied on semantic analysis of texts. It's able to rank easily the difference between «he did a tremendous good work» or «we have been pleased to work with him» and «he's a good guy to work with».

AI then bring an aseptic analysis and equality on the evaluation even considering the cultural difference each country may have, which is relevant to Wec (Worldwide ethics committee) Mevaluate Holding, which elaborates the «country–notes» to help understand the score expressed by the «digitalized reputational rating».

2.2 Foresee Cultural Changes

Mevaluate Holding «digitalized reputational rating» is time-weighted, meaning that it evolves as time passes by. This is not only related to new facts added but to the idea that every fact is evaluated from a historical perspective.

AI must take care of this concept analyzing historical patterns for facts throughout all the ranked «population» and suggesting fine-tuning actions on algorithm weights. An example may be useful.

In the early '60 in western countries, when evaluating a manager to be hired by a company, the marital status was as much relevant as the military career. A divorce or a separation from his wife could have led to a bad evaluation.

Nowadays divorces and separations are so frequent that these facts are, in most cases, no more relevant to a manager evaluation for hiring.

AI helps to foresee trends not only for a specific fact but also finding out correlations among types of fact suggesting Wec Mevaluate Holding intervention on weights that determine the score expressed by the «digitalized reputational rating». AI helps in keeping the algorithm and the resulting ratings updated and meaningful.

3 Case Study-Anbsc: «Digital Antimafia»

An application case (in the economic and judicial field) of AI related to the score expressed by the «digitalized reputational rating» in Italy concerns the so-called «Digital Antimafia» developed in the context of the Public Private Partnership (case study) between:

- National agency for the administration and destination of seized and confiscated assets from organized crime (Anbsc), supervised by the Ministry of the Interior (Italian Government),
- Professional association with traceable reputation auditors (Apart), founded by Mevaluate Holding Ltd, supervised by the Ministry of Economic Development (Italian Government) pursuant to italian law 4/2013,
- Crop News (personalized objective reputational reports), online magazine published by Crop News Onlus (founded by Mevaluate Holding Ltd),
- Consumers and Users Federation (Codacons, Codici, Konsumer Italia) and
- National judicial administrators institute (Inag).

Prevention Measures (art. 34 and 34-bis of italian legislative decree number 159/2011)—guidelines for preventive compliance for companies and entities—use the «digitalized reputational rating» for the construction of «management and control organizational model» to construct the exemption from liability for administrative offenses related to facts dependent on crime.

The economic crisis is a pernicious influence that breaks down the «immune defenses» of companies. Therefore, it is necessary to find a «prophylaxis», which allows to prevent and face the risk of «diffusion», implementing an efficient «management and control organizational model». This should be the instrument that promptly identifies the risks to which the company is exposed, and intervenes incisively, in order to avoid any «contagion».

The Italian system of anti-mafia prevention is now considered—in Europe, but also worldwide—as the most advanced and effective legislation model for contrasting high-profitability organized crime—not only more mafia-style. Inevitably, it poses a problem of «sustainability» in periods of unfavorable economic conditions and/or in a business environment affected by endemic structural deficits.
The economic crisis can generate:

- a greater vulnerability of businesses: the economic crisis amplifies the need for liquidity of companies and, fatally, the likelihood to resort to money of illicit origin and
- a greater possibility of mafia infiltration: organized crime, can therefore succeed in establishing itself more easily within the most vulnerable enterprises.

The Italian law of 17 October 2017, n. 161 modified the codicist system (Italian legislative decree number 159/2011 so-called «antimafia code»), introducing new «non-ablative» (so-called «gentle») instruments (art. 34-bis italian legislative decree number 159/2011—judicial control) and modernizing existing ones (art. 34 italian legislative decree number 159/2011—judicial administration of «contaminated» companies), for the purpose of legality remediation, that is specifically,

- strengthening the anti-mafia prevention system, with the need to safeguard business freedom and maintain employment levels,
- definitely restraining the settlement of the criminal phenomenon in the business environment, without foreseeing an anticipated "exclusion" of the subjects entitled to the management of the company subject to the judicial procedure.

A particularly innovative aspect concerns the recipient subjects of «antimafia disqualification» (art. 84, paragraphs 3 and 4, of italian legislative decree number 159/2011). In fact, provided they have challenged the relative provision issued by the Prefect, these subjects may ask the Regional Administrative Court to access the «judicial control» (article 34-bis, paragraph 6, italian legislative decree number 159/2011).

How can companies prevent the «trauma» of the corporate system at risk of criminal/mafia infiltration? Adopting or revising the «management and control organizational model» with a view to the «criminal» liability of the company (better, administrative liability arising from a crime committed, in the interest of the same company, by management and/or employees) pursuant to italian legislative decree number 231/2001, which consists of:

- analyzing (with the possible help of sector specialists) the existing «management and control organizational model»,
- analyzing and implement existing procedures in the company to make them comply with the parameters established by italian legislative decree number 231/2001,
- adopting (in the FORM and in the SUBSTANCE) a new «management and control organizational model» compliant with the provisions of italian legislative decree n. 231/2001 and proper company procedures and
- appointing an adequate «supervisory body» (in Italy «Organismo di Vigilanza», so-called OdV), providing it with appropriate independent control powers (and related expense funds).

AI applied to the «digitalized reputational rating»—developed by Mevaluate Holding and published by the online magazine Crop News—represents a formidable tool in the preparation and implementation of the above mentioned «Model 231/2001», if applied:

- to support risk management systems,
- to support the evaluation of counterparties (suppliers, customers, business partners, collaborators, etc.),
- to qualify and be distinguished towards counterparties,
- to support the construction of exempting account from responsibility in any criminal proceedings pursuant to italian legislative decree number 231/2001,

and, at the same time, it represents an added value for the company and/or institution, distinguishing it in terms of reliability and transparency.

AI and Law

Processing of Personal Data and AI: GDPR Guarantees and Limits (Between Individual Data and BIG DATA)

Angela Busacca and Melchiorre A. Monaca

Abstract The use of AI systems and related applications records a continuous and constant increase in different fields of research and daily life, based on large amounts of data flows that are uploaded and circulating online and, as stated by doctrine, constitute the main resource of the digital economy. In recent years, however, the development of communication channels and information flows has led to an exponential increase in the amount of data available, making it possible to develop a series of AI applications. On these considerations it is necessary to identify the legal framework of the relationship between circulation online and processing of data and AI systems, both with reference to personal data (which constitute the most significant percentage) and with reference to non-personal data, regarding to the regulatory framework existing (based mainly on the GDPR and Regulation UE 1087/2018). If, indeed, the spread and increase of AI systems can contribute to the EU strategy of increase the digital single market, it is nevertheless necessary to maintain the optimal standards of protection of personal data and, more generally, of the protection of the personality rights *on* (and *in*) the Net.

Keywords Artificial Intelligence · Law · Privacy · GDPR · Big data

1 Introduction. AI Challenges and Legal Categories

The development of new technologies and the spread of informatics have, since the beginning, posed many questions from a legal point of view, above all the persistence and validity of traditional legal categories and of current regulatory systems, that are conceived and structured for an eminently physical and material reality (Passaglia and Poletti 2017).

DiGIES—Università Mediterranea di Reggio Calabria: this paper is the result of the common studies and dialogue of the two authors; in the drafting of the final text, Angela Busacca wrote paragraphs 1 and 3 and Melchiorre Monaca paragraph 2.

A. Busacca · M. A. Monaca (✉)
Mediterranea University of Reggio Calabria, 89124 Reggio Calabria, Italy
e-mail: melchiorre.monaca@unirc.it

D. Marino and M. A. Monaca (eds.), *Economic and Policy Implications of Artificial Intelligence*, Studies in Systems, Decision and Control 288,
https://doi.org/10.1007/978-3-030-45340-4_6

In particular, since the mid-1990s, with the massive spread of Internet and the possibility for millions of users to enter and interact on the Internet, the same questions have focused on this new reality, projecting people, in an increasingly pervasive way, into a virtual dimension next to the real (and tangible) dimension, thereby determining new declinations of legal categories, of subjective rights and of (juridical) relationships between them. Law and jurisprudence have to face renewed needs and new questions, balanced between an evolutionary interpretation able to adapt the current rules to new realities and the creation of an ad hoc system, only just conceived and specifically addressed to the de-materialized reality of Internet.

In particular, the starting point is the new consideration of "data" and "information", which, from the traditional configuration in terms of legal and economic assets, takes on the value of "system measurement unit" and "key value" of the Net Economy (Gambino et al. 2019; Pascuzzi 2017; Sica et al. 2016). Indeed, Internet lives in the data flows that both private and professional cyber-users input, create, share and exchange; on Internet, perhaps, this freedom of online data circulation has been repeatedly indicated as a cardinal principle of every set of rules and every intersection between self-regulation and state-regulation (Poletti and Causarano 2019).

These same questions are renewed today, regarding those situations that are commonly referred to as "the challenges of artificial intelligence (AI)", which today represent the first of the critical points of a continuous and frenetic technological evolution (Contucci 2019; Italiano 2019), characterized by a collapse of times and by an acceleration in front of which legal operators and the legislator himself sometimes appear unprepared.

In the context of the different options offered by the *polisemia* of terms, AI concerns a complexity of studies and activities that allow to "create machines capable of solving problems and offering services and activities of their own, independent of human intelligence " in an ICT ecosystem. On the basis of this definition, AI projects, as stated by the doctrine, into a dimension of trans-humanism (Ruffolo and Amidei 2019; Messinetti 2019) and re-writes the relationship between man and machine in terms of "thinking subject-thinking object", "creator and creature" but in a dynamic mode, involving all the fields of human sciences.

So as to avoid every possible misunderstanding, it seems appropriate to underline immediately that, in terms of law and in the context of traditional legal categories, AI systems fall fully within the scope of the discipline on the juridical object, in the evolutionary interpretation of art. 810 of the Civil Code (things that can be object of rights are reinterpreted, rereading "things" as "entities" and therefore also immaterial, de-materialized and digital goods). An AI system does not escape from its "object" scope of application to rise to a non-human subjectivity that would represent an apocalyptic scenario but rather poses a new series of questions on the correct interpretation to be attributed to the categories of book III of the Civil Code. AI poses new questions to define the legal status of some new phenomena that are difficult to relate to the mainstream of traditional categories, first of all, the relationship between digital goods, which could hardly be placed in the concepts of compound things, accessories, relevance or universality.

Besides the renewed reflection on book III c.c. categories (Gorassini 2018), however, there are many areas of reflection that emerge: from the profiles related to contracts (in the dimension of the so-called smart contracts and the applicability of the discipline of book IV cc), to regulation for the so called company 4.0 (with all the reflections on the subject of freedom of economic initiative, organization and protection of business activities and the work of book V c.c), to the different topics of responsibility and protection of rights deriving from the use of AI applications in different areas, from medical diagnostics to predictive activities, both in terms of prevention and evaluation of damages (with references to the disciplines contained in book IV and in book VI cc).

In addition to the intersection in all the areas of economic relevance, likewise, also possible inference with the personal sphere of subjects cannot be neglected, meaning as "subject" every user acting as a developer, programmer and service recipient of the AI systems performance: in the physical dimension, with the possibility of subrogating parts of the body with "smart" mechanical tools (prostheses and artificial limbs) to the immaterial dimension that includes the different declinations of the human personality and the legal situations that derive thereof.

"Immaterial attributes of the personality", to use the words of the doctrine (Zeno Zencovich 1993), are those that are immediately affected by the development of technological systems (whether they develop linear algorithms or AI systems of machine learning or deep learning): "datification" exposes physical subjects to the potential risks of increasingly invasive technologies; projecting himself on the Net, the cyber-user interacts and generates continuous flows of information, so the AI systems can, through data collection and processing activities, deconstruct and reconstruct the most intimate and sensitive areas of a human being (Pizzetti 2018; Orefice 2018).

It is precisely because of these considerations that particular attention must be paid to the activities of data collection and processing, regarding personal and non-personal data, voluntarily uploaded into the electronic circuit or unintentionally made available for processing activities that are not always linear and sequential; for instance, just think of the use of everyday smart devices or the most frequent and widespread home automation applications. Due to connection relationships, a data flow is generated that circulates within an AI system and on which activities can be performed—processing, structuring and construction of data patterns and calculations—but also AI activities that allow systems to create new data (information from information: the Big Data formula) and implement the "knowledge" of the machine to reach, through structured neural networks and so called learning methods (such as "trial and error"), predictive assessments and (increasingly) autonomous choices (D'Acquisto 2018; Messinetti 2019).

In relation to the protection of data flows, particularly those that, in circular and bi-univocal ways, have fed the circuits of knowledge and of commerce since the development of Web 2.0, the attention of legislators (both in a national dimension and in the EU) and legal operators had a dual purpose: on the one hand, to protect and promote the free circulation of personal and non-personal data in a functional perspective of economic exchanges; on the other hand, particularly in recent years, to guarantee high standards of protection especially for personal data (particularly for those considered "sensitive") and offer suitable guarantees against any abusive or hidden activities that could harm or cause damage to the data subject (that is to which the same data refer).

The evolution matured by the legislator in terms of the right to data protection, with a passage that the doctrine has identified as "from a functional right" to a "fundamental right", is reflected in the succession of European measures, developing from the instrument of Directive (dir95/46/CE) to Regulation (Reg. 679/2016 GDPR) and identifying a (higher) standard of unified protection and a renewed series of obligations for the subjects who carry out the data processing activities, with a view to guaranteeing the data subject against any critical issues and pathological situations relating to data activities (Di Resta 2018; Finocchiaro 2017; Sica et al. 2016; Bolognino et al. 2016).

The GDPR has reiterated the principle of free circulation of online data, but has also made consent of the data subject as the "centre of gravity" of the system of circulation of personal data, with an approach that, although it appears valid and even optimal for all processing activities with algorithms put in place with the (classical) architectures of Von Neumann, it may appear instead insufficient if referring to data mining methods and to the development of neural networks that provide a hidden layer articulation for subsequent and hidden treatments that cannot be predetermined when the data is collected and the (relative) consent given.

Precisely in consideration of this specific topic and respecting the role of the present contribution to the Conference work, the following considerations will be addressed to offer a brief examination of some critical issues that AI activities (with particular attention to data mining and the structuring of so-called Big Data, as well as the development of neural networks that implement machine learning methods) can generate in reference to processing activities and protection of personal data, particularly in the case of sensitive data, in relation to both collection and archiving methods and the possibility of processing carried out by subjects other than the original controller and not foreseen/foreseeable at the time of data collection. On this basis, we will try to evaluate what the possible solutions offered by the GDPR are, even if in doubt, as indicated by part of the doctrine, of a substantial insufficiency of the regulatory apparatus, conceived on an individual consent model, but maybe limited in cases of dimensions and volumes of macroscopic data and processing activities, subsequently, and in exponential development.

2 The Algorithm Society, AI Systems and (Market) "value" of Data

With regard to the processing and protection of personal data, the starting point must be a reflection on the object of the activities themselves, i.e. the different categories of data. This paper will analyses only personal data (as defined by art. 4 GDPR[1]), and thus it seems appropriate to underline how the increase of communication flows, the use (more and more massive) of mobile devices and the development of IoT connections have determined a quantitative increase of data uploaded and circulating online (Falce et al. 2017; Zeno Zencovich 2018).

Online data categories include personal and non-personal data (for the protection of which, moreover, the EU has recently approved Regulation 1087/2018), which can be structured data, taken from a previous catalogue in databases and subsequently extracted for further processing activities or, more commonly, unstructured data, collected through social media profiles or communication and interaction channels (for example chat, forums, e-commerce sites, YouTube channels) to which the data subject has given consent, probably by filling in the special form with authorization for processing activities pursuant to art. 13 GDPR.

Leaving aside, for the moment, questions concerning the value and awareness of such consent, especially in relation to "subsequent" processing activities and to the real situations of information asymmetry between controller and data subject, what is currently relevant is the increase in the amount of data that can be subjected to processing activities and that constitute a considerable resource from which it is possible to extract statistics, knowledge and results so as to create an economic advantage on the market (Contucci 2019; Naldi 2018; Moro Visconti 2018).

Data analysis and various activities that the GDPR defines as "processing",[2] with an illustrative and non-exhaustive meaning, allow to obtain not only immediate results but also to be able to obtain further "information from information", thus determining the phenomenon commonly referred to as Big Data.

The chronicle of recent decades has shown that the traditional centripetal consideration of the right to privacy as a right to non-interference in one's own personal sphere was already outdated with the advent of the first computers and more traditional and "linear" computations, making place for a recognition of the right to protection of

---

[1] Art. 4 defines, for the purpose of Regulation, personal data as "any information relating to an identified or identifiable natural person ('data subject'); an identifiable natural person is one who can be identified, directly or indirectly, in particular by reference to an identifier such as a name, an identification number, location data, an online identifier or to one or more factors specific to the physical, physiological, genetic, mental, economic, cultural or social identity of that natural person".

[2] Art. 4 defines, for the purpose of Regulation, processing activities as any operation or set of operations which is performed on personal data or on sets of personal data, whether or not by automated means, such as collection, recording, organization, structuring, storage, adaptation or alteration, retrieval, consultation, use, disclosure by transmission, dissemination or otherwise making available, alignment or combination, restriction, erasure or destruction.

personal data, such as the right to knowledge and the control of information flows uploaded online or stored in databases.

Since the advent of Web 2.0 and with all the developments that today lead us to speak of the IV industrial revolution of the 4.0 Company, the right to protect personal data must be proposed in a declination of ever greater inclusion and sharing, becoming aware of the ineluctable development of processing activities and of the role they play both for the purposes of public bodies (and the realization of common interests) and for the (more individualistic) purposes of private companies (Finocchiaro 2019; Tosi 2019; Bichi 2019).

The GDPR translates all these instances into the centrality of the data subject's consent and in the enforcement of the principle of minimization and principle of purpose (specified, explicit and legitimate), identified as the cornerstones of the lawfulness of the processing and which, already present in Directive 95/46/EC are re-proposed in the light of the new principle of accountability (Bassini et al. 2018).

Processing activities, in a nutshell, are carried out through computation systems and algorithms that allow to reach a series of outputs in conformity to the problems and to the questions proposed, according to settings and instructions given by the programmer/developer and aimed at acquiring a specific result.

From these first considerations, it seems clear that the collection and analysis of unstructured data, potentially acquired from social media pages or from stored information on a chat or a YouTube channel, allow processing activities to extract information of a varied and different *nature*. This heterogeneous set of information, which will be subsequently further processed and aggregated, generates a consequently varied and different *value* for the market of tangible goods and, in parallel, for what we could define the "market of ideas and opinions", correlated to the public sphere and social life.

The circulation and system of analysis algorithms therefore acquire, in turn, an economic value on the market, and the exchange value of the processed data patterns is increased due to the newly acquired greater correspondence to the capacity to show ways and strategies for better positioning of assets on the market or orientation of public opinion. On this basis, neither the definition of today as an "algorithm society" nor reflections of the doctrine that fears a "dictatorship of the algorithm", with impact on lives and individual choices, appear random (Zellini 2018; Rulli 2018).

Up to this point, we have referred to a quantitative dimension of the data and of the increases recorded. Indeed, growth in terms of volume and mass of data, moreover, constitutes one of the distinctive features of the three-dimensional growth that has led to the phenomenon of Big Data (Volume, velocity and variety), generically indicated as a metaphor for developments in the ICT society and which probably represents the end-point of contemporary "datification".

In the definitions that have been provided, from the first one in 2011 to the present, Big Data indicates a complex, articulated set of data, elaborated and structured through traditional algorithmic techniques and databases but also, always with greater frequency and multiple modes, through AI systems, including machine learning and deep learning, NoSQL and MapReduce.

While it may seem simplistic to define Big Data only as a set of structured and complex data and information, the characteristics of the phenomenon become more evident by analysing the characteristics, referred to as the 5 Vs: velocity, variety, volume, veracity and value. Big Data indicates a series of complex data structures that are created, managed, processed and transferred in extremely (and increasingly) reduced times (velocity); that are drawn from a heterogeneous series of sources and determine, for this very extreme heterogeneity and diversification, multiple articulations and implications that represent different moments and areas of life and individual choices (variety); that, precisely because they are acquired through different sources (data base, web channels, IoT, smart sensors and devices), increase and articulate more and more (volume); that, for the purposes of efficiency of the output result, must have a sufficient degree of reliability and correspondence to truth (veracity) and that, ultimately, acquire, precisely because of the premises to offer a result, in terms of analysis or predictive evaluation or result, the possibility to represent a competitive advantage on the market and an economic asset (value).

These characteristics also reflect the evolution of the methods and tools of analysis applied: it is necessary to highlight that only the first three (velocity, variety and volume) are found in the original Laney model of 2001, while the other two emerge from a subsequent study of 2015, in which criteria and elements are also evaluated (veracity and value), thanks to AI tools and machine learning methods used, which are able to condition the final results (Ottolia 2017; De Gregorio and Torino 2019).

The use of AI systems determines the characterization of the "information from information" phenomenon: Big Data appears as the result of a process of extraction from the inexhaustible mass of data present on the Net, "processed" through complex procedures, that can be based on predetermined deterministic algorithms or on the evolutionary and cognitive processes of applied AI systems (D'Acquisto and Naldi 2017; Bifulco 2018).

Transfer of data involves the possibility of carrying out new, different processing activities, also by a controller other than the one who originally acquired (or extracted the data). Furthermore, the application of AI modalities determines a requalification also in a qualitative sense of data patterns, which can be correlated, composed and analysed in exponential mode. The data mining methods applied to Big Data (Big Data Analytics) can then determine significant results regarding the study and assessments of individual and collective choices, even providing guidance keys for public opinion or mass consumerism.

To obtain a valid result, however, an assessment in terms of the quantity of data collected and processed is no longer sufficient: the element that acquires value is the reliability of the data, that is its correspondence to a real and truthful situation; in the presence of an unreliable figure, in fact, the final result would be irretrievably altered, losing value (Musacchio et al 2018).

The use of AI systems, therefore, involves greater attention to the quality of the data, as well as to quantity, and, consequently, also to the origins and sources of the data itself; likewise, as also highlighted by art. 29 Working Party, the use of AI systems involves a greater percentage of risk to which the data subject is exposed, both in terms of invasion of privacy (especially in the case of data disclosed unintentionally, unstructured and present, sometimes unconsciously, in sources on which the AI agents draw) both in terms of protection of personal data and control of circulation (especially in the case of data, entered on the network with consent but subsequently transferred and structured by other controllers).

3 "Information from Information". AI, BIG DATA and Personal Data Between Processing and Protection: Which Kind of Regulation?

On the basis of the considerations made, it is clear that the quantitative and qualitative increase in personal data flows submitted to processing activities concerns a greater risk index for the data subject and therefore requires the adoption of optimal protection measures able to offer adequate remedial tools (both technical and legal) in the case of a data breach and, in more general terms, of damage in the legal sphere of the data subject.

The European legislator, precisely on the basis of the need to strengthen the protection of data flows and situations regarding the data subject, has designed a new legal framework of reference with the GDPR, aimed at protecting natural persons with reference to personal data processing and free data flow circulation on the Net. For these purposes, GDPR provides the implementation of technical, organizational and legal measures but also a renewed and greater consideration of the principles of privacy by design and privacy by default (art. 25[3]) as well as the rights of the data subject in the case of automated processing (Article 22[4]).

[3]Art. 25 Data protection by design and by default: Taking into account the state of the art, the cost of implementation and the nature, scope, context and purposes of processing as well as the risks of varying likelihood and severity for rights and freedoms of natural persons posed by the processing, the controller shall, both at the time of the determination of the means for processing and at the time of the processing itself, implement appropriate technical and organizational measures, such as pseudonymization, which are designed to implement data-protection principles, such as data minimization, in an effective manner and to integrate the necessary safeguards into the processing in order to meet the requirements of this Regulation and protect the rights of data subjects. 2. The controller shall implement appropriate technical and organizational measures for ensuring that, by default, only personal data which are necessary for each specific purpose of the processing are processed. That obligation applies to the amount of personal data collected, the extent of their processing, the period of their storage and their accessibility. In particular, such measures shall ensure that by default personal data are not made accessible without the individual's intervention to an indefinite number of natural persons. 3. An approved certification mechanism pursuant to Article 42 may be used as an element to demonstrate compliance with the requirements set out in paragraphs 1 and 2 of this Article.

[4]Art. 22 Automated individual decision-making, including profiling: (1) The data subject shall have the right not to be subject to a decision based solely on automated processing, including profiling, which produces legal effects concerning him or her or similarly significantly affects him or her. (2) Paragraph 1 shall not apply if the decision: (a) is necessary for entering into, or performance of, a contract between the data subject and a data controller; (b) is authorized by Union or Member State law to which the controller is subject and which also lays down suitable measures to safeguard

The very choice of the legal instrument (the regulation) instead of the previous one (the directive) highlights the objective of unification of protection standards and also consideration of the material (including profiling) and territorial (protecting the data of all European citizens also in an international dimension) application areas underlining the passage to strengthened protection; just with the GDPR, as underlined by the doctrine, greater importance regarding data protection is accomplished: from consideration as an instrumental situation, connected to the protection of the market, to consideration in terms of the fundamental right of the natural person (Messinetti 2019; Sica et al. 2016; Finocchiaro 2017).

But yet, despite the undeniable value of the previsions of the GDPR standards, there are some doubts about the hypothesis of data processing with the use of AI systems, particularly in the case of so called "black box" learning machine or deep learning, also with reference to processing that determines the extraction of information from information, creating the phenomenon of Big Data and its exploitation (Pizzetti 2018; Tosi 2019; Finocchiaro 2019).

The European institutions themselves pay constant attention to the AI phenomenon in relation to the protection of personal data (Coppini 2018). In 2018, the European Commission highlighted the need to create a "European data space" functional to the development of AI and based on "trust, data and infrastructure availability" with a view to guaranteeing access to data as "a fundamental element for a competitive AI scenario". These requirements have been reported in Communication COM (2018) 795 which places the need for a reconciliation between data access and data protection in the first instance. Also in 2018, the European Commission for the Efficiency of Justice in the Council of Europe adopted the "European ethical charter on the use of artificial intelligence in judicial systems and in related fields". Subsequently, in 2019, the European Parliament has adopted the Resolution of 12 February 2019, on the subject of the global European industrial policy in the field of robotics and AI, and the need to access and use data for the realization of new activities was confirmed, while guaranteeing adequate standards of protection of rights of data subjects (Farace 2019; Moro Visconti 2019).

The technical reference framework, as can be seen, becomes increasingly complex and it seems appropriate, therefore, to question the validity and applicability of the GDPR legal framework also in Big Data hypotheses and processing activities carried out with AI systems. The emergence of both phenomena on a large scale, and subsequent discussion, took place in the final phase of preparation for the text of the GDPR and in the two years following its enactment. For this reason, some commentators on the doctrine had already expressed doubts and identified some

---

the data subject's rights and freedoms and legitimate interests; or (c) is based on the data subject's explicit consent. (3) In the cases referred to in points (a) and (c) of paragraph 2, the data controller shall implement suitable measures to safeguard the data subject's rights and freedoms and legitimate interests, at least the right to obtain human intervention on the part of the controller, to express his or her point of view and to contest the decision. Decisions referred to in paragraph 2 shall not be based on special categories of personal data referred to in Article 9(1), unless point (a) or (g) of Article 9(2) applies and suitable measures to safeguard the data subject's rights and freedoms and legitimate interests are in place.

interpretative critical points in relation to the processing activities carried out on large amounts of data (and by controllers other than those who collected the data), for which the principles of consent and the principle of purpose, if construed strictly, could prove to be a limit to the development of activities and the creation of new knowledge. Further doubts concerned rules on the subject of automated decisions and processing pursuant to Article 22, on the dual point of the configurability of a so called "right of explanation" for the data subject and of the traceability of the title of controller (and related liability rules) in the case of use of an AI application or of a robotic system.

It is possible to mention here only a few further notes on the problems indicated, in particular on that relating to the protection of the rights of the data subject in the case of automated decisions (pursuant to Article 22), in relation to which GDPR supersedes EU Convention 108/1981, revised in 2018 and indicated today as Convention 108+.

Reading the text of art. 22 GDPR, in combination with the previous art. 15 paragraph Ilett. h),[5] two distinct situations can be identified: on the one hand, the prohibition of decisions based only on the use of automated systems, and on the other hand, the right of the data subject to access, in the case of processing activities pursuant to art. 22, information on the activities, the underlying logic, as well as the importance and expected consequences (for the data subject himself) (Belisario 2018; Caia 2018; Finocchiaro 2017).

Using automated computing systems for the processing of personal data is therefore not questioned, rather the prohibition refers only to the hypothesis of decisions based only on automated processing. Along with the possibility of using AI systems and automated methods (for a part of the processing, up to the extreme of a decision based on a chain of processing activities of which only one is not automated), there arises, as a point of balance between the rights of the controller (to use the data to process them with AI application) and of the data subject, the right to obtain information about the underlying logic and (possible) consequences (Pellecchia 2019). This is a legal situation subject to analysis and debate in doctrine, since according to some it would be possible to define a real "right to explanation" (also on the basis of recital n.71 which prefigures a right to "obtain human intervention, express one's opinion, obtain an explanation of the decision obtained and contest it"; Messinetti 2019), while for others it would only be an access to information on specific points and modality of the AI systems used (as foreseen by art. 15; Finocchiaro 2019); we would be in the presence of a sort of antinomy between the wider situation prefigured in the recital and the more limited right of access provided, instead, by Article 15.

In the case of automated decisions, taken with the use of AI systems, protection of the rights of the data subject appears to be limited to the possibility of access to information on the processing methods, the logic used and the possible consequences;

---

[5]Article 15 Right of access by the data subject: (1) The data subject shall have the right to obtain from the controller confirmation as to whether or not personal data concerning him or her are being processed, and, where that is the case, access to the personal data and the following information: … (h) | the existence of automated decision-making, including profiling, referred to in Article 22(1) and (4) and, at least in those cases, meaningful information about the logic involved, as well as the significance and the envisaged consequences of such processing for the data subject.

but also from this consideration some critical points emerge, because this protection appears addressed, in essence, to lead the data subject to obtain a series of information, which should make him aware of the processing activities carried out on his data and fill a situation that we could define as "information asymmetry", referring to a term widely used in the field of contracts between parties with unequal bargaining power. Recalling the term and using it in this context does not appear out of place: also in this case the controller and the data subject are in an imbalanced position with respect to the knowledge of the processing activities on personal data; and, considering the data as the fee for a whole series of services offered (apparently) without monetary consideration, a comparison with contract law is quite appropriate.

However, a doubt arises as to the real value of the right of access to information on the processing activities and the underlying logic as a valid form of protection for the data subject, given that the information to which he would have access to could prove to be difficult to understand as it relates to eminently technique aspects. Furthermore, there is another question relating to the controller (or more generally the subject) to whom to address the request, given that in the case of transfer of data patterns for subsequent processing in view of the final decision, the controller who collected the data (as well as the initial consent to the treatment) could be a subject different from the one that then carries out the (successive) processing for the purposes of the automated decision. The data subject, in this case, should identify the controller (or the controllers) to whom to address his/her requests and against whom to put forward any liability suit. Also on this last point, it should be emphasized that, even in the case of automated processing and damage caused by AI applications, it does not appear entirely clear who the interlocutor is to be summoned to assert any suits of liability: certainly it cannot be the AI system, considered as a subject (as also hypothesized by the Resolution of the EU Parliament of 16 February 2017, relative to recommendations on civil law rules applicable to robotics, in terms of "specific legal status for robots" or "electronic-personalities"; Busto 2017). The need to identify a subject to whom to make the claim, in terms of liability rules in the case of damage and for any compensation, should then be directed towards the user, that means the controller who profits from the application AI or, as envisaged by some, to the programmer or application developer.

The considered point is, however, only one of the critical points of the discipline of the GDPR, in its possible applications to the use of AI application and systems, but it highlights the necessity of a rethinking of some of the GDPR proposed categories, addressed to a relational dimension controller-data subject, to adapt them to the different dimensions of Big Data and the use of AI systems.

To reach a conclusion, which is actually the proposition of new study and research itineraries, we can say that the GDPR constitutes the essential starting point for every evaluation in terms of protection of personal data, but doubts cannot be silenced regarding the validity of a regulation conceived and articulated in an individual dimension, for an application aimed, instead, at large masses of data that pose different problems, above all those related to the reliability of the data and to the identification of a (possible) limit in the data mining methods and AI tools, that can be used lawfully. The 108+ Convention (CONVENTION 108 1981), as amended in

2018, can offer valid support, especially with reference to all those activities that fall within the category of automated decisions, and the Code of Ethics on the use of AI is a good starting point for a more general discourse of compatibility between technological development and fundamental rights. However, the gap remains in a regulation that goes beyond the individual horizon, looking at data protection no longer as a personal situation, but as the right of the community, underlying a wider interest that goes beyond individuals to take on a universal scope.

Data security, understood as security of circulation and guarantee of the rights of the data subject, could then assume the value of a common good and, as such, require protection as a fundamental right and as a social right, the interest of the community and a point of attention for national and European legislators.

The intervention of the European legislator on AI is the most desirable solution, with a view to integration and not to replacing the GDPR, which would in any case remain valid, and represents the real challenge of these years, to respond to the acceleration of technology and the sense of disorientation that often seizes the (unprepared) legal operators in the face of sudden innovations.

However, the question remains about the real ability of legislators to meet this challenge. Above all to collect it in good time.

References

Bassini, M., Liguori, L., Pollicino, O.: Sistemi di Intelligenza Artificiale, responsabilità e accountability. Verso nuovi paradigmi? In: Pizzetti, F. (a cura di) Intelligenza artificiale, protezione dei dati personali e regolazione, p. 333. Ed. Giappichelli Torino (2018)

Belisario, E.: art. 15. Diritto di accesso dell'interessato. In Riccio, G.M., Scorza, G., Belisario, E. (a cura di) GDPR e normativa privacy, p. 219. Ed. WKI, Milano (2018)

Bichi, R.: Intelligenza artificiale tra "calcolabilità" del diritto e tutela dei diritti in Giurisprudenza Italiana, p. 1772 (2019)

Bifulco, R.: Intelligenza artificiale, internet e ordine spontaneo. In Pizzetti, F. (a cura di) Intelligenza artificiale, protezione dei dati personali e regolazione, p. 383. Ed. Giappichelli Torino (2018)

Bolognino, L., Pelino, E., Bistolfi, C.: Il nuovo regolamento privacy europeo. Ed. Giuffrè Milano (2016)

Busto, N.: La personalità elettronica dei robot: logiche di gestione del rischio tra trasparenza e fiducia. In Cyberspazio e diritto, p. 499 (2017)

Caia, A.: art. 22 Processo decisionale automatizzato relativo alle persone fisiche, compresa la profilazione. In Riccio, G.M., Scorza, G., Belisario, E (a cura di) GDPR e normativa privacy, p. 219. Ed. WKI, Milano (2018)

Contucci, P.: Intelligenza artificiale tra rischi ed opportunità in Il Mulino, p. 637 (2019)

Coppini, L.: Robotica e intelligenza artificiale: questioni di responsabilità civile in Politica del diritto, p. 713 (2018)

D'Acquisto, G.: Qualità dei dati e Intelligenza Artificiale: intelligenza dei dati e intelligenza dei dati. In Pizzetti, F. (a cura di) Intelligenza artificiale, protezione dei dati personali e regolazione, p. 265. Ed. Giappichelli Torino (2018)

D'Acquisto, G., Naldi, M.: Big Data e Privacy by Design. Ed. Giappichelli Torino (2017)

De Gregorio, G., Torino, R.: Privacy protezione dei dati personali e Big Data. In Tosi, E. (a cura di) Privacy Digitale, p. 447. Ed. Giuffrè, Milano (2019)

Di Resta, F.: La nuova "privacy europea". Ed. Giappichelli Torino (2018)

Falce, V., Ghidini, G., Oliveri, G.: Informazione e Big data tra innovazione e concorrenza. Ed. Giuffrè Milano (2017).
Farace, D.: Privacy by design e privacy by default. In Tosi, E. (a cura di) Privacy Digitale, p. 485. Ed. Giuffrè, Milano (2019)
Finocchiaro, G.: Intelligenza artificiale e protezione dei dati personali. In In Giurisprudenza Italiana, p. 1670 (2019)
Finocchiaro, G.: Il nuovo regolamento europeo sulla privacy e sulla protezione dei dati personali. Ed. Zanichelli Bologna (2017)
Gambino, A.M., Stazi, A., Mula, D.: Diritto dell'informatica e della comunicazione. Ed. Giappichelli Torino (2019)
Gorassini, A.: Lo spazio digitale come oggetto di un diritto reale? In Rivista di diritto dei media (2018)
Italiano, G.F.: Le sfide interdisciplinari dell'intelligenza artificiale. In Analisi giuridica dell'economia, p. 9 (2019)
Messinetti, R.: La tutela della persona umana versus l'intelligenza artificiale. Potere decisionale dell'apparato tecnologico e diritto alla spiegazione della decisione automatizzata in Contratto e Impresa, p. 861 (2019)
Moro Visconti, R.: L'intelligenza artificiale: modelli di business e criteri di valutazione in Il diritto Industriale, p. 421 (2018)
Moro Visconti, R.: La valutazione delle blockchain: Internet of Value, network digitali e smart transaction, in Il diritto Industriale, p. 301 (2019)
Musacchio, M., Guaita, G., Ozzello, A., Pellegrini, M.A., Ponzani, P., Zilich, R., De Micheli, A.: Intelligenza Artificiale e Big Data in ambito medico: prospettive, opportunità e criticità. J. AMD, **21-3**, 204 (2018)
Naldi, M.: Prospettive economiche dell'Intelligenza Artificiale. In Pizzetti, F. (a cura di) Intelligenza artificiale, protezione dei dati personali e regolazione, p. 225. Ed. Giappichelli Torino (2018)
Orefice, M.: I Big Data e gli effetti su privacy, trasparenza e iniziativa economica. Ed. Aracne Roma (2018)
Ottolia, A.: Big data e innovazione computazionale. Ed. Giappichelli Torino (2017)
Pascuzzi, G.: Il diritto dell'era digitale. Ed. Il Mulino Bologna (2017)
Passaglia, P., Poletti, D.: Nodi virtuali legami informali: internet alla ricerca di regole. Pisa University press (2017)
Pellecchia, E.: Privacy, decisioni automatizzate e algoritmi. In Tosi, E. (a cura di) Privacy Digitale, p. 417. Ed. Giuffrè, Milano (2019)
Pizzetti, F.: La protezione dei dati personali e la sfida dell'Intelligenza Artificiale. In Pizzetti, F. (a cura di) Intelligenza artificiale, protezione dei dati personali e regolazione, p. 5. Ed. Giappichelli Torino (2018)
Poletti, D., Causarano, M.C.: Autoregolamentazione privata e tutela dei dati personali: tra codici di condotta e meccanismi di certificazione. In Tosi, E. (a cura di) Privacy Digitale, p. 369. Ed. Giuffrè, Milano (2019)
Ruffolo, U., Amidei, A.: Intelligenza artificiale e diritti della persona: le frontiere del "transumanesimo". In Giurisprudenza Italiana, p. 1658 (2019)
Rulli, E.: Giustizia predittiva, intelligenza artificiale e modelli probabilistici. Chi ha paura degli algoritmi. In Analisi Giuridica dell'Economia, p. 533 (2018)
Sica, S., D'Antonio, V., Riccio, G.M.: La nuova disciplina europea della privacy. Ed. Giappichelli Torino (2016)
Tosi, E.: Privacy digitale, persona e mercato: tutela della riservatezza e protezione dei dati personali alla luce del GDPR e del nuovo Codice della Privacy. In Tosi, E. (a cura di) Privacy Digitale, p. 1. Ed. Giuffrè, Milano (2019)
Zellini, P.: La dittatura del calcolo. Ed. Adelphi Milano (2018)
Zeno Zencovich, V.: Dati, grandi dati, dati granulari e la nuova epistemologia del giurista. In Rivista di diritto dei media (2018)

Zeno Zencovich, V.: Profili negoziali degli attributi della personalità, in Il diritto dell'informazione e dell'informatica, p. 545 (1993)
CONVENTION 108—28 January 1981—Council of Europe Convention for protection of Individuals with regards to Automatic Processing of Personal Data
Modernised Convention for protection of Individuals with regards to Automatic Processing of Personal Data—18 May 2018—Council of Europe, 128th Session of the Committee of Minister
GUIDELINES ON AI AND Data PROTECTION (T-PD 2019/01), Consultative Committee of the Convention on the Protection of Individuals with regard to Automatic Processing of Personal Data (Convention 108)
REGULATION (EU) 2018/1807 OF THE EUROPEAN PARLIAMENT AND OF THE COUNCIL of 14 November 2018 on a framework for the free flow of non-personal data in the European Union
REGULATION (EU) 2016/679 OF THE EUROPEAN PARLIAMENT AND OF THE COUNCIL of 27 April 2016 on the protection of natural persons with regard to the processing of personal data and on the free movement of such data, and repealing Directive 95/46/EC (General Data Protection Regulation)
COMUNICATION 795/2018 Communication from the Commission to the European Parliament, the European Parliament, the European Council, the Council, the European economic and social Committee, and the Committee of the Regions on a "Coordinated Plan on Artificial Intelligence"
CEPEJ European Ethical Charter on the use of artificial intelligence (AI) in judicial systems and their environment, adopted at the 31st plenary meeting of the CEPEJ (Strasbourg, 3–4 December 2018)
European Parliament Resolution 12 February 2019 on a comprehensive European industrial policy on artificial intelligence and robotics 2018/2088 (INI)

Algorithmic Bosses and What to Do About Them: Automation, Artificial Intelligence and Labour Protection

Valerio De Stefano

Abstract This paper aims at filling some gaps in the mainstream debate on automation and the future of work. This debate has concentrated, so far, on how many jobs will be lost as a consequence of technological innovation. This paper examines instead issues related to the quality of jobs in future labour markets. It addresses the detrimental effects on workers of awarding legal capacity and rights and obligation to robots. It examines the implications of practices such as People Analytics and the use of big data and artificial intelligence to manage the workforce. It stresses on an oft-neglected feature of the contract of employment, namely the fact that it vests the employer with authority and managerial prerogatives over workers. It points out that a vital function of labour law is to limit these authority and prerogatives to protect the human dignity of workers. It then highlights the benefits of human-rights based approaches to labour regulation to protect workers' privacy against invasive electronic monitoring. It concludes by highlighting the crucial role of collective regulation and social partners in governing automation and the impact of technology at the workplace. It stresses that collective dismissal regulation and the involvement of workers' representatives in managing and preventing job losses is crucial and that collective actors should actively participate in the governance of technology-enhanced management systems, to ensure a vital "human-in-command" approach.

Keywords Artificial intelligence · Workplace surveillance · People analytics · Management-by-algorithm · Labour Law

---

BOF-ZAP Professor of Labour Law at KU Leuven, the University of Leuven. This chapter draws on the article *'Negotiating the Algorithm': Automation, Artificial Intelligence and Labour Protection* published in a special issue of the *Comparative Labor & Policy Journal* on "*Automation, Artificial Intelligence, and Labour Protection*" guest-edited by me (41 Comp. Labor Law Policy J. (2019)). This chapter and the special issue were published within the framework of the Odysseus grant "Employment rights and labour protection in the on-demand economy" that I received from the *FWO Research Foundation—Flanders*.

V. De Stefano (✉)
KU Leuven, Leuven, Belgium
e-mail: valerio.destefano@kuleuven.be

D. Marino and M. A. Monaca (eds.), *Economic and Policy Implications of Artificial Intelligence*, Studies in Systems, Decision and Control 288,
https://doi.org/10.1007/978-3-030-45340-4_7

1 Introduction

In 2013, the publication of a now world-famous paper by Oxford scholars Carl Frey and Michael Osborne on "how susceptible jobs are to computerisation" spurred a gargantuan debate on automation and the related threats of unrestrained job losses and mass unemployment.[1] Arguably, this debate focused overwhelmingly on the number of jobs that will be lost to automation. Indeed, the academic and policy debate on these issues has largely adopted a "quantitative" approach, trying to estimate the number of workers that could be put out of a job as a consequence of technologic breakthroughs.[2] Some studies have criticized these estimates, pointing out some of their possible flaws and also concentrating on the potential benefits of technological progress in terms of job creation.[3] So far, however, this debate has not sufficiently considered the qualitative aspects connected with job automation. Much less attention, in fact, has been devoted to the quality of the jobs that will remain, but that will require growing interactions between humans and technological tools, both in the forms of advanced machinery and of software used to manage businesses and production processes.[4]

It almost seems taken for granted that these "jobs of the future" will require high technical skills, that new machinery and programmes, complemented by artificial intelligence, will absorb routine, menial and dangerous tasks and that the fortunate workers who remain employed will have access to highly rewarding jobs, with technology playing a liberating role for them. Therefore, according to this view, instead of focusing on the quality of these jobs, regulators should be concerned in making sure that the highest number of persons possible acquire the skills necessary to be employed in these liberated roles; they should also envisage measures to absorb occupational shocks determined by automation and to mitigate its social consequences for workers that will be displaced and will not be able to develop these high-level skills or will not find employment because there will be fewer jobs available.[5]

This narrative, however, follows a techno-deterministic approach that should be called into question. To begin with, it assumes that technological breakthroughs will always imply progress, particularly for the fortunate workers who have developed the skills to remain in employment after the introduction of new machinery and business processes. This assumption, however, risks proving excessively optimistic. While it is probably true that technology will be able to automate some routine and unpleasant tasks, it will also increase the possibility of management to increasingly monitor working activities in a way that is not desirable for worker.[6] Software and hardware are already spreading in modern workplaces that allow management to

[1]Frey and Osborn (2013).

[2]See, for instance, the well-known paper of Frey and Osborn (2013). For an in-depth discussion on manufacturing processes, see Dauth et al. (2017).

[3]The literature on the topic is already enormous, see Autor (2015), OECD (2016, 2018). *See, for a general critical discussion*, Kucera (2017). For an in-depth legal discussion, *see* Estlund (2018).

[4]An exception is Eurofound (2018).

[5]McKinsey Global Institute (2017).

[6]*See below* Sect. 2.

give workers instructions on the work they do and to control their performance through digital tools.[7] Artificial intelligence, the use big data and "management-by-algorithm" are already a reality in the world of work,[8] potentially leading to very intrusive business practices. The risks connected to these practices are almost absent from the mainstream debate on the future of work and on the effects of automation, even if, as argued below, the introduction of advanced machinery in the workplace can materially spur these risks.

Another assumption that follows this techno-deterministic approach is that these developments are inevitable—in other terms, they are the price to pay to benefit from the rewards of technological progress. Accordingly, limiting the functioning of new technologies at the workplace would inescapably reduce progress for economies and societies at large, supposing that these limits could theoretically be imposed through regulation. Moreover, the mainstream narrative on automation also risks leading to the impression that regulation over the introduction of new technological tools and machinery and their implications on the quantity and quality of jobs cannot be put in place and that any attempt to govern the effects of technological breakthroughs would hamper innovation and lead to economic losses.

These assumptions must all be questioned. Regulation aimed at mitigating the potentially detrimental effects of the use of technological devices on job quality and workers' human dignity already exists in various countries of the world. Moreover, many jurisdictions already have in place regulation aimed at mitigating the social impact of mass redundancies and job losses, also connected to automation and technological innovation. A detrimental economic impact from this regulation has not been proved. On the contrary, strong involvement of social partners and regulators in the management of potential mass redundancies is associated with high levels of productivity and innovation, in addition to the benefits for workers.[9]

Most importantly, regulation is also fundamental in governing how automation and the introduction of new technologies will impact on the quality of the jobs that will be affected by them rather than merely focusing on their quantity. Labour legislation and collective bargaining must play a much more central role if these phenomena are to take place in a way that respects the human dignity and the fundamental rights of workers—yet, these aspects are still under-researched in the vast debate on automation and the future of work. This contribution wants to fill some of these gaps in this debate. The next Section starts doing so by indicating how some technological innovations can lead to intrusive managerial practices that could magnify these risks.

---

[7] Akhtar et al. (2018). See also the articles of Aloisi and Gramano (2019).

[8] Pasquale (2015), Dagnino (2017).

[9] *See below*, Sect. 4.

2 Technologically-Enhanced Workers' Monitoring: Artificial Intelligence, Big-Data and the Risks of Algorithmic Discrimination

Technological tools and digitalised supervision systems are increasingly used to manage the workforce in modern workplaces.[10] Workers' surveillance is, of course, nothing new; business historians such as David Landes have long reported that concentration of workers in factories started occurring before mechanisation, to surveil and direct the workforce better than what was possible in processes based on dispersed homework.[11] Fordist-Taylorist business models were also based on extensive monitoring of workers.[12]

Information technology and artificial intelligence,[13] however, allow monitoring workers activities to extents unthinkable in the past, as well as the gathering and processing of an enormous amount of data on these activities.[14] More and more workers, for instance, use wearable work instruments that enable registering of their movements and location minute by minute, also measuring their work pace as well as breaks. Data collected through wearables, including sociometric badges,[15]are often analysed using artificial intelligence to assess workers' productivity and fitness to execute particular tasks.[16] Wearables are also used or experimented in warehouses and other workplaces to direct workers to their next assignment. Goods in Amazon warehouses, for instance, are stored apparently at random. Amazon workers are guided by technological tools to the next item to pick and process, a system that also enables the company to automatically track and measure the speed and efficiency of every individual worker. Workers who underperform according to the metrics of the automated surveillance systems can receive warnings or see their employment terminated automatically "without input from supervisors".[17]

---

[10]Moore et al. (2018), Ajunwa et al. (2017).

[11]Landes (1969).

[12]Stone (2004).

[13]The term "artificial intelligence", in this paper, is used as a reference to the so-called "narrow artificial intelligence" or "weak artificial intelligence", namely the artificial intelligence used to performed a single task, such as—as a commonly used description goes—"playing chess or Go, making purchase suggestions, sales predictions and weather forecast". This is the only type of artificial intelligence that exists, nowadays. Even self-driving cars are considered merely a sum of several narrow AIs, and the same applies to online translation engines. Narrow AI is commonly opposed to "General AI", i.e. "the type of Artificial Intelligence that can understand and reason its environment as a human would", which has not been developed yet. The direct citations are from Dickson (2017). For a broader discussion of the distinction between "strong" and "weak" AI, see Kaplan (2016).

[14]Emanuele Dagnino, *supra* note 8.

[15]*See below in this Section.*

[16]Pav Akhtar, Phoebe Moore, & Martin Upchurch, *supra* note 7; Manokha (2017).

[17]Lecher (2019). The article reports: "Amazon says supervisors are able to override the process". See also Baraniuk (2015).

GPS systems allow monitoring the position and speed of truck and van drivers as well as of delivery riders and ride-sharing drivers working for on-demand platforms. These systems can also be used to verify, for instance, if these workers gather in specific locations, to prevent or react to collective action.[18] Similar to workers in a warehouse that use automated systems of direction, platform workers are assigned to the next task by the app's algorithms, which are also designed to measure the speed and diligence of the worker in completing the tasks, including by factoring in the rating and reviews that customers assign to workers. Bad scores or performance below the algorithm's standards can lead to the exclusion of the worker from the platform and thus to "dismissal", also made easier by the purported self-employment status of these workers.[19] And this is not confined to tasks "on-the-road". Workers on online "freelancing marketplaces" and domestic workers who are contracted on platforms to do work in customers' households live in constant worry over ratings and how the platforms' algorithms take ratings into account when assigning the next job.[20]

The way these management systems operate is rarely transparent, as companies do not share the methods through which ratings and customers' feedbacks over the workers' activities are gathered and processed. Management by the rating is also spreading ever more beyond platform work, with apps that allow processing patrons' and restaurants' feedbacks over individual waiters.[21]

Nor should it be assumed that increased forms of surveillance are confined to low-wage or blue-collar jobs. HR practices that make resort to forms of artificial intelligence that facilitate "management-by-algorithm" and "electronic performance monitoring" are also extensively used in white-collar occupations. Electronic performance monitoring (EPM) has been described by Phoebe Moore et al. as including "email monitoring, phone tapping, tracking computer content and usage times, video monitoring and GPS tracking". According to these researchers, "data produced can be used as productivity indicators; indication of employees' location; email usage; website browsing; printer use; telephone use; even tone of voice and physical movement during conversation".[22] These data, coupled with the use of "big data" analytical instruments, also constitute the basis of so-called People Analytics practices. Pioneering legal studies on this topic, conducted by Matthew Bodie, Miriam Cherry et al. define "People Analytics" as:

> a process or method of human resources management based on the use of "big data" to capture insights about job performance. The core idea is that unstructured subjective judgment is not rigorous or trustworthy as a way to assess talent or create human resources policies. Instead, data—large pools of objective, generally quantitative data—should form the foundation for decision-making in the HR space.[23]

---

[18] De Stefano (2016).

[19] Aloisi (2016).

[20] Foundation for European Progressive Studies (FEPS) (2017).

[21] O'Donovan (2018), Fillon (2018).

[22] Pav Akhtar, Phoebe Moore & Martin Upchurch, *supra* note 7.

[23] Bodie et al. (2017).

Data are therefore collected from a vast array of sources.[24] One of the companies at the forefront of these practices, *Humanize*, reports on its webpage that metadata can be obtained from "email and call timestamps, number of chat messages sent, and duration of meetings can be measured to uncover patterns on how teams actually work". This does not necessarily mean that the actual content of messages and chats is examined, as the company claims to include "no names or content in the metadata".[25]

Nonetheless, even if these individual-content data are not collected or are effectively anonymised, collection practices can be highly invasive and aimed at detecting highly personal elements,[26] including the level of interaction with colleagues and even the humour of workers, for instance through the use of so-called "sociometric badges". These are wearable devices that allow monitoring the location of workers, their movements and also, through the use of incorporated microphones and voice-pitches analysis the mood of workers without actually recording the content of their conversations.[27]

EPM is also being used to monitor workers in telework and smart work arrangements, which allow workers to perform their activities outside of traditional workplaces, and are thus usually associated with higher workers' autonomy.[28] Companies like Crossover sell systems such as the Worksmart Productivity Tool to monitor teleworkers and other remote workers by taking screenshot of their computers at fixed intervals and collecting additional data, including, as the company's website explains: "keyboard activity, application usage, screenshots, and webcam photos to generate a timecard every 10 min". This timecard is then shared with the workers and their managers via a "logbook where all of your timecards are displayed and a dashboard summarizes your timecards to show you how you spent your time".[29] Other companies market web-filtering software, like Interguard, that record and reports on data such as web history and bandwidth utilization "whether the employee is on or off network".[30]

All these data can also be processed through AI tools that rate workers on various performance metrics. In 2019, for instance, the Guardian reported that dozens of firms in the United Kingdom, including several law firms, employed AI to scrutinize staff

---

[24]For a thorough review carried out by a public authority of common EPM practices see Article 29 Data Protection Working Party (now, the European Data Protection Board), *Opinion 2/2017 on data processing at work*, adopted on 8 June 2017.

[25]Humanize, https://www.humanyze.com.

[26]According to the Article 29 Data Protection Working Party (now, the European Data Protection Board), *supra* note 24: "The risk is not limited to the analysis of the content of communications. Thus, the analysis of metadata about a person might allow for an equally privacy-invasive detailed monitoring of an individual's life and behavioural patterns".

[27]Fischbach et al. (2009).

[28]*The workplace of the Future*, The Economist, Mar 28, 2018, https://www.economist.com/news/leaders/21739658-artificial-intelligence-pushes-beyond-tech-industry-work-could-become-faireror-more; Solon (2017).

[29]Crossover, https://www.crossover.com/worksmart/#worksmart-productivity-tool.

[30]Interguard, https://interguardsoftware.com/web-filtering.html.

behaviour, also to identify "influencers" and "change-makers" in the workforce.[31] Interestingly, this practice is not so new. Cathy O'Neill discussed the case of a company, Cataphora, which in 2008 marketed a system to identify "idea generators" in the workforce by analysing corporate emails and messaging. When the 2008 recession hit, HR managers began to lay-off people starting by those who performed poorly under Cataphora's metrics. As O'Neill, a mathematician and data scientist, explains these programs risk, among other things, to be highly inaccurate since they are based on limited data.[32]

Business-sponsored wellness programs also use software like Fitbit to track employees' fitness.[33] This, among other things, can contribute to having access to information related to off-duty activities of workers. Surveillance of workers' off-duty activities is also nothing new, suffice here to think of the Social Department of Ford,[34] which famously investigated the lifestyles of workers in the motor company. However, the blurring of boundaries between work and life, the constant interconnection with IT devices and digital services such as social networks and technological devices that allow to gather data from individuals' online and offline conducts makes it possible to accede to a flow and amount of information that is very difficult to quantify and limit in advance. Articles in the press also reported cases of monitoring practices that aimed to prevent fraud by snooping social network activities and statuses.[35]

Personal data gathered on the Internet, also by acceding to information available through social networks is also increasingly used to make hiring decisions,[36] and the practice of asking employees to disclose their social network passwords is also spreading, so that 18 individual states of the United States passed legislation explicitly banning it.[37]

People Analytics and EPM, of course, can sometimes be rooted in genuine business needs such as fostering productivity and raising levels of security, also to the benefit of individual employees. Wearables that analyse fitness data, for instance, can be employed to mitigate health and safety risks, including stress, and to prevent accidents.[38] Workers may also be interested in using systems that help them staying focused on their jobs both when they are on-site and off-site and having their activities recorded accurately so that—if anything goes amiss—they can prove to have acted diligently. Business and workers can also be interested in the prevention of illicit behaviours such as fraud as well as forms harassment that can occur online. Moreover, HR practices such as People Analytics are also grounded in the idea that artificial intelligence can help better manage the workforce by eliminating individual

[31] Boot (2019).

[32] O'Neill (2016).

[33] Ifeoma Ajunwa, Kate Crawford & Jason Schultz, *supra* note 10.

[34] Matthew T. Bodie, Miriam A. Cherry, Marcia L. McCormick & Jintong Tang, *supra* note 23.

[35] Olivia Solon, *supra* note 28.

[36] Emanuele Dagnino, *supra* note 8.

[37] Matthew T. Bodie, Miriam A. Cherry, Marcia L. McCormick & Jintong Tang, *supra* note 23.

[38] *The workplace of the Future*, *supra* note 28.

biases of supervisors and replacing them with more objective and neutral metrics.[39] The use of artificial intelligence and other technological tools to supervise working activities, therefore, should not be regarded as necessarily negative.

The practices discussed above, however, can also lead to very severe intrusion into workers' private life and materially infringe their privacy,[40] by allowing management to access to very intimate information, including, for instance, through the use of data based on medical insurance claims on the intention to become pregnant and on the possibility to develop sickness.[41] Wearables and security cameras, programs that register online and offline activity, as well as take screenshots of computers, can also turn into extenuating practices of endless surveillance. Far from fostering workforce performance, these models can also generate stress as well as adverse reactions and cause sharp declines in efficiency and productivity.[42]

In addition to this, the idea that management-by-algorithm and artificial intelligence can necessarily lead to more objective and bias-free HR practices may prove substantially wrong. The risk is that these systems reflect the biases of their human programmers and only focus on their ideas around productivity and work performance, for instance by discarding or penalising job candidates or workers with disabilities or with features that differ from the expectations programmers have. The scarcity of diversity in tech companies can also exacerbate these phenomena. In an official Opinion on artificial intelligence, the European Economic and Social Committee recently observed: "the development of AI is currently taking place within a homogenous environment principally consisting of young, white men, with the result that (whether intentionally or unintentionally) cultural and gender disparities are being embedded in AI, among other things because AI systems learn from training data". The Committee warned against the misconception that data is by definition objective. Data, instead, "is easy to manipulate, may be biased, may reflect cultural, gender and other prejudices and preferences and may contain errors".[43]

The risk, therefore, is that management-by-algorithm and artificial intelligence at the workplace, long from having neutral outcomes and reducing discrimination, could augment discriminatory practices.[44] A vast literature already exists that highlights how algorithm-based decision-making can perpetuate discriminatory practices and marginalisation of vulnerable groups, especially when the collection of data is poor.[45] This form of decision-making is often based on data that reflect past behaviours.[46] If those behaviours were biased, the likelihood that any automated-decision process propagates those biases in the future is very high.[47] Imagine a system of automatic

[39] Matthew T. Bodie, Miriam A. Cherry, Marcia L. McCormick & Jintong Tang, *supra* note 23.

[40] Hendrickx (2015).

[41] Ifeoma Ajunwa, Kate Crawford & Jason Schultz, *supra* note 10.

[42] Pav Akhtar, Phoebe Moore & Martin Upchurch, *supra* note 7.

[43] European Economic and Social Committee (2017).

[44] Matthew T. Bodie, Miriam A. Cherry, Marcia L. McCormick & Jintong Tang, *supra* note 23.

[45] Frank Pasquale, *supra* note 8; Noble (2018).

[46] Cathy O'Neill, *supra* note 32.

[47] Eubanks (2018).

scanning of CVs for hiring or promotion. If this system is built on data about the previous hiring in the company or sector, there is a high chance that it can mimic past recruitment practices. If, in turn, those practices were discriminatory or skewed, they could be perpetuated in the future and, what is worst, this would occur under an "aura" of perceived objectivity usually credited to machines. Nor would it be simple to remove discrimination by merely instructing the algorithms to ignore sensitive data such as gender or race, since sophisticated software could still recognize, and penalize, subjects underrepresented in the previous hiring on the basis of other data. For instance it could use certain types of career breaks and as proxies to recognize women or postcodes or first and last names to identify members of minorities. This risk is even more severe when these practices are based on self-learning artificial intelligence, with software being able to reprogram their own criteria and metrics to reach a very general predefined outcome, such as improving work productivity. The lack of transparency and the risk of dehumanizing work would then be even more exacerbated.

Nor it should it be taken for granted that a one-dimensional vision of productivity and efficiency embedded into artificial intelligence technologies would necessarily lead to better business outcomes. Algorithms are often being used to implement just-in-time work practices that scale the workforce's figures and shifts by the expected business demand, thus contributing to a casualization of work patterns and job and income instability that goes far beyond the "usual suspects" in the platform economy. A study conducted by various universities on retail workers, for instance, shows that algorithms aimed at fostering business' efficiency can lead to suboptimal results, as a consequence of these algorithms being based on a very limited notion of efficiency and therefore not be taking into account the many hidden costs associated with schedule instability.[48]

One oft-overlooked dimension of advanced forms of automation is its potential role in introducing technology-enhanced management of workers facilitated by artificial intelligence. A smart-robot is, in the definition proposed by the EU parliament report discussed at Sect. 2, a robot that has the "capacity to acquire autonomy through sensors and/or by exchanging data with its environment (inter-connectivity) and the analysis of those data" and the "capacity to adapt its behaviours and actions to its environment". Robots that collect the personal data of employees, including by measuring their biological data through interaction with fitness applications and wearables, to enhance productivity or attune the pace or other features of the work to the particular conditions of workers are not impossible to introduce. This is particularly true for co-bots, which, as discussed above, are by definition meant at having a direct physical interaction with human beings and at sharing workspaces with workers.

Moreover, the use of artificial intelligence, management-by-algorithm and People Analytics are, per se, a form of automation of middle-managerial and managerial roles. Managing and disciplining platform workers via workers' ratings is arguably a way of outsourcing assessment of work performance to customers facilitated by

---

[48] Adler-Milstein et al. (2018). See also the discussion of automated scheduling in Berg (2019).

algorithms.[49] EPM has also the potential to increasingly automate core business functions such as HR and also displace the associated clerical occupations, adding to the list of professionals that can be severely affected by automation, together with lawyers and medical doctors.[50]

The implications of these managerial practices, therefore, warrant serious attention by policymakers and scholars and the consequences on privacy, diversity, employment as well as business productivity should be carefully assessed. Even the most well-intentioned measures, including wellness programs, risk turning into forms of dystopian and paternalist control unless a serious reflection on the use of technology at the workplace is carried out.

The paternalism behind EPM is well represented in this statement from the CEO of Awareness Technology, the company that markets *Interguard*, a monitoring system for on-site and remote workers: "if you are a parent and you have a teenage son or daughter coming home late and not doing their homework you might wonder what they are doing. It's the same as employees."[51]

Comparing employees to underage son and daughters is nothing new. In discussing privacy and employers' managerial prerogatives at the workplace, Matthew Finkin recalls that in 1884 the Tennessee Supreme Court did not object to an employer telling employees where to shop—as a father could order his children where to buy goods, so could employers to their employees.[52] Beyond the irony of finding ancient arguments somehow replicated in the most cutting-edge work scenarios, the possibility of management unduly and excessively compressing workers' autonomy and privacy is a structural feature of the contract of employment.[53] As scholars Bodie, Cherry et al. point out, unless regulation specifically limits managerial prerogatives, "in the workplace, there is no legal protection against surveillance per se [...]. The need for monitoring follows from our legal conception of employment, which is based on control: an employee is one whose work is controlled by her employer" and it is the right of employers to specifically direct employees activities "that separates employees from independent contractors".[54]

---

[49] Valerio De Stefano, *supra* note 18.

[50] See also Jerry Kaplan, *supra* note 13.

[51] *Cited by* Olivia Solon, *supra* note 28.

[52] Finkin (2017).

[53] Hendrickx (2014).

[54] Matthew T. Bodie, Miriam A. Cherry, Marcia L. McCormick & Jintong Tang, *supra* note 23.

3 The Importance of Labour and Human Right Protection in Governing Technology at Work

The policy and journalistic discussions on automation have also stirred an extensive debate on universal basic income (UBI).[55] Numerous tech entrepreneurs and companies have maintained that one of the responses to the displacement of jobs caused by automation should be the introduction of UBI, to mitigate the social impact of mass technological unemployment.[56] The debate on UBI is broader than, and goes beyond, these proposals. Several labour advocates have suggested UBI as a progressive policy that would help to face significant challenges in modern labour markets, including technological unemployment and the growth of casualised and unstable forms of employment.[57] This is a very complicated issue that cannot be treated here.[58] What is important to state, however, is that even if a functioning UBI scheme were possible to implement, this would not affect the legal structure of employment contracts and regulation discussed above.

Neoliberal proponents of UBI often take for granted that this measure would substitute for other welfare schemes, including social security. A corollary of this vision is also that, if a UBI were introduced, employment regulation could be rolled back because, in system where everybody had a secure access to income, regulation aimed at supporting workers' income and remediate against their weak bargaining position would no longer be needed, also because the UBI would likely increase their reservation wages.[59]

These assumptions are in line with conventional accounts of employment regulation and mainstream approaches to employment policy. Indeed, the objective of the flexicurity approach to employment the protection is to replace protection of workers "on the job" with protection "on the market", by deregulating aspects of employment protection while securing workers' income through unemployment benefits and active labour market policies.[60]

Policies aimed at substituting protection of employment rights for protection of income risk neglecting an essential feature of employment regulation, which is not just safeguarding workers because they are economically dependent on their employers and have weak bargaining power "on the market", but is also limiting and rationalizing the unilateral exercise of managerial prerogatives "on the job", i.e. while they are employed.[61]

---

[55]Romano and Zitelli (2017).

[56]Sadowski (2016).

[57]*See, for instance*, Standing (2004), Hollo (2016).

[58]See, however, Rogers (2019) and the other articles dealing with UBI published in that same Journal's issue.

[59]Zwolinski (2014). Janine Berg, *supra* note 48, also dismisses the idea that a UBI could adequately substitute for employment protection.

[60]Sciarra (2007).

[61]De Stefano (2014).

Regulation against discrimination, working time regulation protecting the physical and mental health of workers against the risks of fatigue and burnout, rules protecting privacy at the workplace against abusive forms of monitoring, to cite only some of the regulation that limits the exercise of managerial prerogatives cannot be swapped with protection "on the market". This regulation concerns powers and duties that are functioning during the entire course of the employment relationship and do not merely depend on the superior bargaining power of employers but are also enshrined in legal norms. The idea of replacing labour protection at the workplace with securing the stability of income neglects fundamental aspects of the employment relationship, which warrant regulatory limits aimed at protecting human dignity at the workplace. This is also something to take into account when discussing the possibility of introducing UBI or any other form of income protection—even if UBI schemes were introduced, there would still be need of employment regulation and labour protection "on the job".

The fundamental features of employment regulation and its ambivalence in granting far-reaching and intensive unilateral managerial powers that can materially compress the workers' autonomy, on the one hand, and limiting and rationalising those powers, on the other hand, must be particularly heeded in the wake of automation and the increasing use of technological tools to direct the workforce. EPM, People Analytics and the use of artificial intelligence and big data at the workplace magnify the possibility of supervising workers and closely monitoring the performance of working activities. As already discussed in Sect. 2, these technologies can enable egregiously invasive practices and lead to arbitrary and discriminatory outcomes. Indeed, these practices can lead to a "genetic variation" of managerial prerogatives, by "upgrading" them to levels unheard of in the past. Constant attention must thus be paid to these developments and regulation is all the more needed to prevent managerial abuses that imperil the human dignity of workers.

To this end, it is also essential to frame workers' rights in fundamental and human rights discourses. The nature of labour rights as human rights has long been debated[62] and it has also been enshrined in a vast number of international treaties and sources of law.[63] One of the rationales to recognise labour rights as human rights lies precisely on the existence of managerial prerogatives. As discussed above, legal systems vest employers with authority over their workforce that goes beyond social norms and is underpinned by legislation. Limiting and rationalising authority to preserve human dignity—which is one of the essential functions of human rights—is also essential at the workplace.[64] Labour protection, by limiting the exercise of managerial prerogatives, is also crucial to ensure that the authority of employers is not exerted in ways that jeopardise the human rights of workers.

---

[62] Fenwick and Novitz (2010), Arthurs (2006), Mantouvalou (2012).

[63] Politakis (2007).

[64] For an extensive discussion of how protection of the human dignity and human rights of workers can be posed as a foundational element of labour law, see the contributions collected in Philosophical Foundations of Labour Law (Hugh Collins, Gillian Lester, and Virginia Mantouvalou eds. 2019). For an in-depth critical appraisal of human-rights based arguments in labour-law discourses, see, however, Finkin (2019).

Human rights approach to labour regulation can indeed prove beneficial also concerning the protection of workers' autonomy and dignity regarding electronic monitoring of their activities.[65] The European Court of Human Rights, for instance, has interpreted the right to private life under article 8 of the European Convention on Human Rights to enshrine the protection of privacy of individuals at the workplace. In a case that concerned the dismissal of a worker for the use of the internet at work for private purposes, in a situation where the employer had access to the content of the workers' communications via IT tools, the Court established that employers' monitoring of online activities, while admissible in principle, had to be carried out proportionately, to ensure that arbitrariness and abuses be avoided.[66] Among the safeguards that the Member States have to consider, to determine whether monitoring practices are legitimate, the Court indicated: the circumstance that employees be properly notified of the possibility that the employer might monitor correspondence and other communication; the presence of legitimate reasons to justify monitoring the communications and accessing their content; the possibility to establish less intrusive monitoring practices. The Court also mandated to consider, in general, the extent of the monitoring and the degree of intrusion into the workers' privacy, making a distinction between access to the metadata covering the flow of communications and access to the content of these communications.

This judgment can provide a general protective framework for workplace relations in countries that adhere to the European Convention on Human Rights of the Council of Europe. Notably, the Council of Europe also recently updated its Convention for the Protection of Individuals with regard to Automatic Processing of Personal Data. The new text of the Convention, after its entry into force, will provide for an individuals' right "not to be subject to a decision significantly affecting him or her based solely on an automated processing of data without having his or her views taken into consideration".[67]

For countries that also belong to the European Union, further guidance can be found in the General Data Protection Regulation (GDPR). The GDPR, however, is no panacea in itself against the excesses of management-by-algorithm and the use of AI at the workplace. Firstly, commentators noted that EU law has been interpreted by the Court of Justice of the EU (CJEU) as to provide for lower protection in case a decision is taken based on subjective inferences drawn on data rather than on objective and verifiable facts.[68] This is a paradox, considering the possible detrimental impacts that wrong inferences can cause—imagine if a decision on hiring or promotion is made by inferring how somebody with a particular credit history can perform into an employment contract, without taking into account what factors contributed to that credit history.

Also, the CJEU has so far refused to extend the remit of EU data protection law to the accuracy of decision-making processes. And even the GDPR provisions that

[65] See Hendrickx (2019).

[66] *Bărbulescu v. Romania*, No 61496/08, ECHR 2017.

[67] Article 9 of the revised Convention.

[68] Mittelstadt and Wachter (2019).

seem to provide more meaningful protection in this area could prove insufficient. For instance, Art 22(1) of the GDPR grants for the right not to be subject to "a decision based solely on automated processing", when this decision produces legal or "similarly significant[...]" effects.[69] Most likely, however, a high number of decisions concerning workplace issues will fall into the exceptions to this rule allowed by Art 22(2), being they "necessary for entering into, or performance of, a contract".[70] In this case, the GDPR mandates that employers or other data controllers implement "suitable measures to safeguard the data subject's rights and freedoms and legitimate interests, at least the right to obtain human intervention on the part of the controller, to express his or her point of view and to contest the decision". Workers, therefore, will have the right to contest fully automated decisions that affect them significantly. This protection, however, will be vain unless they can show that a specific "enforceable legal or ethical decision-making standard" has been violated. Without these standards, the protection under Art. 22 risks remaining "an empty shell".[71]

It is, therefore, crucial that adequate and specific standards and protections be provided in the world of work. In this respect, Article 88 of the GDPR is a crucial provision. It provides that the EU Member States may introduce, by law or by collective agreements, "specific rules to ensure the protection of the rights and freedoms in respect of the processing of employees' personal data in the employment context". These rules shall "include suitable and specific measures to safeguard the data subject's human dignity, legitimate interests and fundamental rights" with particular regard to "monitoring systems at the work place", transparency of processing and transfer of personal data.[72]

These regional approaches to workers' privacy protection, founded on the idea of protection of human and fundamental rights at the workplace, and specifically addressing the need that the prerogatives of managing and monitoring workers do not impinge upon their human dignity, can guide the introduction (or the update) of labour regulation aimed at safeguarding workers against abusive supervision practices in the wake of the spread of technology-enhanced monitoring systems.[73] A human-rights based approach, grounded on the idea that the human right to privacy can

---

[69]For an in-depth account of the potential shortcomings of Article 22, *see* Floridi et al. (2017). A critical question will concern the interpretation of the word "solely" in this context. Adequate standards are needed to ensure that nominal involvement of humans that sanction decisions made by automatic mechanisms will not deprive data subjects of the protection under Art 22.

[70]Another case of exception is when data subjects give their express consent to solely automated decision-making. It is worth noting, however, that the Article 29 Data Protection Working Party (now, the European Data Protection Board) in its *Opinion 2/2017 on data processing at work*, adopted on 8 June 2017, observed: "consent is highly unlikely to be a legal basis for data processing at work, unless employees can refuse without adverse consequences".

[71]Brent Mittelstadt & Sandra Wachter, *supra* note 69.

[72]Article 88, Regulation (EU) 2016/679 of the European Parliament and of the Council of 27 April 2016 on the protection of natural persons with regard to the processing of personal data and on the free movement of such data, and repealing Directive 95/46/EC (General Data Protection Regulation); *see, for initial comments*, Armaroli and Dagnino (2019), Hendrickx (2018), Fusco (2018).

[73]Frank Hendrickx, *supra* note 72.

only be limited insofar as this is indispensable to the exercise of other human rights and that any limitations must be proportionate to this end, can indeed provide a meaningful general framework of protection that may prove beneficial, in contrast to spot-remedy approaches adopted in systems where recognition of workers' rights as fundamental rights is still lagging, like the United States,[74] and to proposals to govern technological innovation based on much vaguer "ethical" principles, such as the currently overhyped "ethical AI" discourse.

A human-rights based approach to labour protection cannot neglect the importance of collective rights such as freedom of association and the right to collective bargaining in the protection of human dignity at the workplace. The function of collective rights is not only to give workers a better position to negotiate economic conditions of employment; collective rights also act as "enabling rights", facilitating securing and effectively enforcing any other right at the workplace. As such, collective rights also serve as a fundamental tool to rationalise and limit the exercise of managerial prerogatives, since they allow counterpoising a collectively organised party to the intrinsic collective and organisational dimension of these employers' prerogatives, which can be exerted on an individual basis but also on the workforce as a whole. Collective rights, including the right to collective bargaining, allow moving from a purely unilateral exercise of those prerogatives towards a consensual governance of work, by requiring negotiations on aspects of the business organisation that would be, in lack of collective relations, unilaterally governed by employers, by means of the authority vested in them by the legal system.[75] Reference to collective bargaining in Article 88 of the GDPR as a mechanism to provide adequate and specific standards in the context of data collection and processing to safeguard the human dignity and the fundamental rights of workers confirms how crucial collective rights are to counter abuses of automated-management practices at the workplace. The next section concludes this Article by exploring how collective regulation is essential to secure adequate labour protection in times of automation and technologically enhanced monitoring practices.

4 "Negotiating the Algorithm": "Human-in-Command" and Collective Rights for the Future of Work

As discussed in the Introduction, the mainstream discourse on automation tends to follow the techno-deterministic assumption that the introduction of new technologies will determine job losses or gains as an autonomous and heterogeneous process impacting labour markets. This approach, nonetheless, does not take into account

[74] For an analysis of the United States' legal framework in this context, see Frank Pasquale, *supra* note 8. See also Ifeoma Ajunwa, Kate Crawford & Jason Schultz, *supra* note 10; Matthew T. Bodie, Miriam A. Cherry, Marcia L. McCormick & Jintong Tang, *supra* note 23; Frank Hendrickx, *supra* note 53.

[75] Liebman (1993).

the role that labour regulation can play to influence this process—something that is indeed surprising, given the high number of international and national instruments that deal with the impact of technology on employment, such as the instruments governing collective dismissals.

Collective dismissals are the subject matter of copious international, regional and national regulation. These instruments commonly require businesses to adequately inform and consult with trade unions and workers' representatives and to involve public bodies before carrying out mass redundancies. Yet having this type of regulation in place is far from sufficient for solving the problems deriving from automation. Job losses could occur at levels unheard-of in the past, for instance, or new technologies could be introduced at a pace that strains current regulation and industrial relations. Moreover, this regulation aims at mitigating the consequences of redundancies but is not able to avert them per se, especially if new machinery and business processes displace a high number of jobs in a short amount of time. Nonetheless, policymakers, researchers and scholars should not start from the assumption that regulation aimed at attenuating mass job losses does not exist or is impossible to apply. Collective redundancies regulation exists, and its existence should be considered when discussing the impact of automation on labour markets, together with the role that social partners and regulators can have in governing these processes.

Nor should it be assumed that regulation would necessarily stifle innovation, another widespread corollary of techno-deterministic approaches to automation. Collective redundancies regulation and labour laws that ensure functioning industrial relations systems and sustain the role of workers' representatives and trade unions can instead be associated with positive economic outcomes.[76] Literature also shows a positive relationship between stronger collective institutions and productivity,[77] economic efficiency, and levels of employment.[78]

The assumption should be, therefore, that collective dismissal regulation and workers' involvement in managing mass redundancies can be beneficial when dealing with automation processes and their social implications.

Moreover, the involvement of workers' representatives can also occur much earlier than when actual redundancies occur. Duties to engage in social dialogue to deal with the envisaged impact of technological innovation are also provided under regional instruments, such as the EU Directive 2002/14.[79] The Directive mandates information and consultation duties both on an ad hoc basis, "on decisions likely to lead to substantial changes in work organisation or in contractual relations" and, on a regular basis, "on the recent and probable development of the undertaking's or the

[76] Adams et al. (2018).

[77] Deakin et al. (2014a), FitzRoy and Kraft (2005).

[78] Deakin et al. (2014b).

[79] Directive 2002/14/EC of the European Parliament and of the Council of 11 March 2002 establishing a general framework for informing and consulting employees in the European Community.

establishment's activities and economic situation". Examples of national regulation that provide for similar duties are also available.[80]

Most importantly, the involvement of workers' representatives can prove particularly beneficial to the aim of governing other implications of new technologies at the workplace, namely those affecting the quality of the jobs that will "survive" after automation. The introduction of artificial intelligence and the use of big data and EPM need to be governed, to ensure that systems that can allow an unprecedented magnification of the scope and impact of managerial prerogatives and the intensity of monitoring do not lead to abuses that impinge on the human rights of workers.

Regulation is needed to govern the amount of data collected on working performances and the personal features of workers, as well as the way data are collected and treated. Nor is this only a matter of privacy protection. The way work is directed through the use of new technologies, including wearables and co-bots among other things, should be regulated to ensure that the quest for higher productivity does not result in occupational hazards and heightened stress for the workers involved. Disciplinary mechanisms facilitated by technology are another key item to regulate. Even if it were possible to have artificial intelligence deciding on issues such as increasing the pace of work or intensifying production, these decisions should always be implemented after a human review. The same goes for any disciplinary measure taken in light of data collected through mechanical monitoring systems or algorithmic processes. Algorithm-based evaluation of work performance should also be disciplined, to make assessment criteria transparent and known to workers and to ensure avoidance of arbitrary or discriminatory outcomes. To this end, again, even if it were possible to have automatic changes and updates in the operation of algorithms through self-learning artificial intelligence, the final decision to amend the criteria through which work performance is assessed should be taken by humans, made transparent and known to workers and also be subject to negotiation.

"Human-in-command", an approach advocated by the European Economic and Social Committee's Opinion on Artificial Intelligence,[81] namely the "precondition that the development of AI be responsible, safe and useful, where machines remain machines and people retain control over these machines at all times" should be strictly followed also concerning work. The Opinion also specifically advocates that "workers must be involved in developing these kinds of complementary AI systems, to ensure that the systems are useable and that the worker still has sufficient autonomy and control (human-in-command), fulfilment and job satisfaction". To fulfil this objective, it is also crucial that any managerial decision suggested by artificial

[80]Swedish Employment (Co-Determination in the Workplace) Act (1976:580), Section 19, for instance, binds employers "to regularly inform an employees' organisation in relation to which [they are] bound by collective bargaining agreement as to the manner in which the business is developing in respect of production and finance and as to the guidelines for personnel policy". Analogous duties are provided also when the employer is not bound by a collective agreement.

[81]European Economic and Social Committee, *Artificial intelligence—The consequences of artificial intelligence on the (digital) single market, production, consumption, employment and society* (own-initiative opinion No. 7, 2017). See now also ILO Global Commission on the Future of Work, Work for a Brighter Future (2019).

intelligence be subject to review by human beings who remain legally accountable, together with their organisation, for the decision and its outcomes. The fact that decisions were taken following machine-based processes should never be a sufficient reason to exclude personal liability; even if electronic personality were introduced in the legal system, humans should always remain accountable for any decision directly affecting workers and any other natural person.

The right not to be subject to fully automated decision-making without human intervention is making its way in supranational regulation. Article 9 of the revised Council of Europe's Convention for the Protection of Individuals with regard to Automatic Processing of Personal Data concerning the right not to be subject to automated decision-making without human intervention, discussed above, together with the provision of the GDPR providing for adequate safeguards in this respect are a step towards establishing a "human-in-command" approach. As argued in the previous section, however, to avoid these provisions remain an empty shell, when it comes to the world of work, specific and adequate standard and regulation are needed in this field.

This regulation will have to remain flexible and quickly adaptable to technological innovation. For this reason, besides a general default legislative framework, detailed and bespoke regulation is essential. In this regard, collective bargaining can play a primary role both at the sectoral and at the workplace level, as recalled in Article 88 of the GDPR. Individual right to access to data and to contest the outcomes of automated decision-making, while essential, could not be sufficient in a context in which technology becomes as pervasive and complex as discussed in the previous Sections. Individuals should not be left alone to cope with the intricacies of this technology when they want to comprehend and contest the consequences of its applications on them.

For this reason, in the world of work, collective rights and voice will be crucial. Collective agreements could address the use of digital technology, data collection and algorithms that direct and discipline the workforce, ensuring transparency, social sustainability and compliance with these practices with regulation. Collective negotiation would also prove pivotal in implementing the "human-in-command" approach at the workplace. Collective bargaining could also regulate issues such as the ownership of the data collected from workers and go as far as creating bilateral or independent bodies that would own and manage some of the data.[82] All this would also be consistent with collective bargaining's fundamental function as an enabling right and as a rationalisation mechanism for the exercise of employers' managerial prerogatives, allowing moving away from a purely unilateral dimension of work governance.

"Negotiating the algorithm" should, therefore, become a central objective of social dialogue and action for employers' and workers' organisation. In 2017, for instance, the UNI Global Union issued a series of cutting-edge proposals on Ethical Artificial

[82]Information and consultation and collective negotiation on data collection and processing are also recommended under the 1997 ILO Code of practice on the protection of workers' personal data. See also Choudary (2018).

Intelligence at the Workplace.[83] Armaroli and Dagnino and Phoebe Moore et al., moreover, report on several collective agreements already in place in various countries that regulate the use of technology not only in monitoring workers but also in directing their work, to protect human dignity and occupational health and safety of workers.[84] In this respect, Seifert also envisages a potentially crucial role for transnational collective bargaining and reports on transnational agreements already concluded on the issue of data protection.[85] Social partners, therefore, are already tackling these issues.[86] Governments also have an essential role to play, in addition to providing a general legislative framework to regulate these issues in lieu of, or complementing, specific collective bargaining. For instance, they can also use fiscal incentives to stimulate technological business strategies on the condition that they fully integrate sustainability objectives and are subject to social dialogue. It will not be a simple process or a quick one, and it will require efforts from all the parties involved. Among other things, substantial resources will need to be spent to ensure that workers, managers, trade unionists and HR personnel be adequately trained to deal with the challenges and opportunities that technology can prompt. Regulation and collective governance of these processes will not be built in a day. However, they are indispensable to ensure that the benefits of technological advancements improve our societies inclusively and as a whole.

References

Adams, Z., Bishop, L., Deakin, S., Fenwick, C., Garzelli, S.M., Rusconi, G.: The economic significance of laws relating to employment protection and different forms of employment: analysis of a panel of 117 countries, 1990–2013. Int. Labour Rev. (2018). https://doi.org/10.1111/ilr.12092

Adler-Milstein, S., Bellisle, D., Fugiel, P.J., Jarpe, M., Kesavan, S., Lambert, S.J., McCorkell, L., Ospina, L.A., Pendem, P., Rapoport, E.D., Williams, J.: Stable scheduling increases productivity and sales. The Stable Scheduling Study. https://worklifelaw.org/projects/stable-scheduling-study/report/ (2018)

Ajunwa, I., Crawford, K., Schultz, J.: Limitless worker surveillance. Calif. Law Rev. **105**, 102 (2017)

Akhtar, P., Moore, P., Upchurch, M.: Digitalisation of work and resistance. In: Moore, P., Upchurch, M., Whittaker, X. (eds.) Humans and Machines at Work: Monitoring, Surveillance and Automation in Contemporary Capitalism (2018)

Aloisi, A.: Commoditized workers: case study research on labour law issues arising from a set of 'on-demand/gig economy' platforms. Comp. Labor Law Policy J. **37**, 653 (2016)

Aloisi, A., Gramano, E.: Artificial intelligence is watching you at work. Digital surveillance, employee monitoring and regulatory issues in the EU context. Comp. Labor Law Policy J. **41** (2019)

---

[83] UNI Global Union (2017).

[84] Phoebe Moore, Martin Upchurch & Xanthe Whittaker, *supra* note 10; Ilaria Armaroli & Emanuele Dagnino, *supra* note 72.

[85] Seifert (2018).

[86] Recently, the OECD also adopted a recommendation calling for social dialogue to play a role about the introduction and use of artificial intelligence at work. See OECD (2019).

Armaroli, I., Dagnino, E.: A seat at the table: negotiating data processing in the workplace. A national case study and comparative insights. Comp. Labor Law Policy J. **41** (2019)
Arthurs, H.: Who's afraid of globalization? Reflections on the future of labour law. In: Craig, J.D.R., Michael Lynk, S. (eds.) Globalization and the Future of Labour Law (2006)
Autor, D.: Why are there still so many jobs? The history and future of workplace automation. J. Econ. Perspect. **29**(3) (2015) (Summer)
Baraniuk, C.: How algorithms run Amazon's warehouse. BBC Future, 18 Aug 2015. https://www.bbc.com/future/story/20150818-how-algorithms-run-amazons-warehouses (2015)
Berg, J.: Protecting workers in the digital age: technology, outsourcing and the growing precariousness of work. Comp. Labor Law Policy J. **41** (2019)
Bodie, M.T., Cherry, M.A., McCormick, M.L., Tang, J.: The law and policy of people analytics. Univ. Colorado Law Rev. **88**, 961 (2017)
Boot, R.: UK businesses using artificial intelligence to monitor staff activity. The Guardian, 7 Apr 2019. https://www.theguardian.com/technology/2019/apr/07/uk-businesses-using-artifical-intelligence-to-monitor-staff-activity (2019)
Choudary, S.P.: The architecture of digital labour platforms: policy recommendations on platform design for worker well-being. ILO Future of Work Research Paper Series, No. 3 (2018)
Dagnino, E.: People analytics: lavoro e tutele al tempo del management tramite big data. Labour Law Issues **3**, 1 (2017)
Dauth, W., Findeisen, S., Südekum, J., Wößner, N.: German robots—the impact of industrial robots on workers. IAB Discussion Paper 30/2017, 2 Oct 2017. https://doku.iab.de/discussionpapers/2017/dp3017.pdf (2017)
De Stefano, V.: A tale of oversimplification and deregulation: the mainstream approach to labour market segmentation and recent responses to the crisis in European countries. Ind. Law J. **43**, 253 (2014)
De Stefano, V.: The rise of the 'just-in-time workforce': on-demand work, crowdwork and labour protection in the 'gig-economy.' Comp. Labor Law Policy J. **37**, 471 (2016)
Deakin, S., Fenwick, C., Sarkar, P.: Labour law and inclusive development: the economic effects of industrial relations laws in middle income countries. In: Schmiegelow, M., Schmiegelow, H. (eds.) Institutional Competition between Common Law and Civil Law: Theory and Policy (2014a)
Deakin, S., Malmberg, J., Sarkar, P.: How do labour laws affect unemployment and the labour share of national income? The experience of six OECD countries, 1970–2010. Int. Labour Rev. **153**, 1 (2014b)
Dickson, B.: What is narrow, general and super artificial intelligence. TechTalks. 12 May 2017. https://bdtechtalks.com/2017/05/12/what-is-narrow-general-and-super-artificial-intelligence/ (2017)
Estlund, C.: What should we do after work? Automation and employment law. Yale Law J. **128**, 254 (2018)
Eubanks, V.: Automating Inequality: How High-Tech Tools Profile, Police, and Punish the Poor. St. Martin's Press, New York (2018)
Eurofound: Game Changing Technologies: Exploring the Impact of Production Processes and Work. Research Report (2018)
European Economic and Social Committee: Artificial intelligence—the consequences of artificial intelligence on the (digital) single market, production, consumption, employment and society (own-initiative opinion), 31 May 2017. JO C 288, 31.8.2017, p. 43 (2017)
Fenwick, C., Novitz, T.: Human Rights at Work: Perspectives on Law and Regulation. Bloomsbury Publishing, London (2010)
Fillon, W.: How rating your server is making their life miserable. Eater, 22 June 2018. https://www.eater.com/2018/6/22/17492528/tablets-restaurants-surveys-score-servers (2018)
Finkin, M.: Chapter 7: privacy and autonomy. Empl. Rights Employ. Policy J. **21**, 589 (2017)
Finkin, M.: Worker rights as human rights: regenerative reconception or rhetorical refuge? In: Bellace, J., ter Haar, B. (eds.) Research Handbook on Labour, Business and Human Rights Law (2019)

Fischbach, K., Gloor, P.A., Lassenius, C., Olguin, D.O., Pentland, A.S., Putzke, J., Schoder, D.: Analyzing the flow of knowledge with sociometric badges. In: COINs. https://www.ickn.org/documents/COINs2009_Fischbach_Gloor_Lassenius_etc.pdf (2009)

FitzRoy, F., Kraft, K.: Co-determination, efficiency and productivity. Br. J. Ind. Relat. **43**, 233 (2005)

Floridi, L., Mittelstadt, B., Wachter, S.: Why a right to explanation of automated decision-making does not exist in the general data protection regulation. Int. Data Priv. Law **7**, 76 (2017)

Foundation for European Progressive Studies (FEPS): Work in the European Gig-Economy (2017)

Frey, C.B., Osborn, M.A.: The future of employment: How susceptible are jobs to computerisation. Oxford Martin School, 17 Sept 2013. https://www.oxfordmartin.ox.ac.uk/downloads/academic/The_Future_of_Employment.pdf (2013)

Fusco, F.: Employee privacy in the context of EU regulation N.2016/679: some comparative remarks. Paper presented at the XVI International Conference in Commemoration of Professor Marco Biagi, Mar 2018

Hendrickx, F.: Employment privacy. In: Blanpain, R. (ed.) Comparative Labour Law and Industrial Relations in Industrialized Market Economies (2014)

Hendrickx, F.: Privacy en elektronisch toezicht. In: Hendrickx, F. (ed.) Arbeidsrecht (2015)

Hendrickx, F.: Privacy, data protection and measuring employee performance. The triggers from technology and smart work. Regulating for Globalization. Trade, Labor and EU Law Perspectives, 21 Mar 2018. https://regulatingforglobalization.com/2018/03/21/privacy-data-protection-and-measuring-employee-performance-the-triggers-from-technology-and-smart-work (2018)

Hendrickx, F.: Article 7. Protection of private and family life and Article 8. Protection of personal data. In: Clauwaert, S., Dorssemont, F., Lörcher, K., Schmitt, M. (eds.) The Charter of Fundamental Rights of the European Union (CFREU) and the Employment Relation (2019)

Hollo, T.: Can less work be more fair? A Discussion Paper on Universal Basic Income and Shorter Working Week. The Green Institute, Canberra City (2016)

Kaplan, J.: Artificial Intelligence: What Everyone Needs to Know. Oxford University Press, Oxford (2016)

Kucera, D.: New automation technologies and job creation and destruction dynamics. ILO Employment Policy Brief, 12 May 2017

Landes, D.: The Unbound Prometheus: Technological Change and Industrial Development in Western Europe from 1750 to the Present. Cambridge University Press, Cambridge (1969)

Lecher, C.: How Amazon automatically tracks and fires warehouse workers for 'productivity'. The Verge, 25 Apr 2019. https://www.theverge.com/2019/4/25/18516004/amazon-warehouse-fulfillment-centers-productivity-firing-terminations (2019)

Liebman, S.: Individuale e collettivo nel contratto di lavoro (1993)

Manokha, I.: Why the rise of wearable tech to monitor employees is worrying. The Independent, 4 Jan 2017. https://www.independent.co.uk/life-style/gadgets-and-tech/why-the-rise-of-wearable-tech-to-monitor-employees-is-worrying-a7508656.html (2017)

Mantouvalou, V.: Are labour rights human rights? Eur. Labour Law J. **3**, 151 (2012)

McKinsey Global Institute: A future that works: automation, employment and productivity, Jan 2017

Mittelstadt, B., Wachter, S.: A right to reasonable interferences: re-thinking data protection law in the age of big data and AI. Columbia Bus. Law Rev. https://ssrn.com/abstract=3248829 (2019)

Moore, P., Upchurch, M., Whittaker, X.: Humans and Machines at Work: Monitoring, Surveillance and Automation in Contemporary Capitalism (2018)

Noble, S.U.: Algorithms of Oppression: How Search Engines Reinforce Racism. New York University Press, New York (2018)

O'Donovan, C.: An invisible rating system at your favorite chain restaurant is costing your server. BuzzFeed News, 21 June 2018

OECD: The risk of automation for jobs in OECD countries: a comparative analysis. OECD Social, Employment and Migration Working Papers, No. 189, 14 May 2016

OECD: Automation, skills use and training. OECD Social, Employment and Migration Working Papers, No. 202, 8 Mar 2018

O'Neill, C.: Weapons of Math Destruction: How Big Data Increases Inequality and Threatens Democracy. Broadway Books, New York (2016)

Organization for Economic Co-operation and Development (OECD): Recommendation of the Council on Artificial Intelligence (2019)

Pasquale, F.: The Black Box Society. The Secret Algorithms That Control Money and Information (2015)

Politakis, G.: Protecting Labour Rights as Human Rights: Present and Future of International Supervision. ILO, Geneva (2007)

Rogers, B.: Basic income and the resilience of social democracy. Comp. Labor Law Policy J. **41** (2019)

Romano, A., Zitelli, A.: Il reddito di base è una cosa seria. Valigia Blu, 7 Mar 2017. https://storie.valigiablu.it/reddito-di-base (2017)

Sadowski, J.: Why Silicon Valley is embracing universal basic income. The Guardian, 22 June 2016. https://www.theguardian.com/technology/2016/jun/22/silicon-valley-universal-basic-income-y-combinator (2016)

Sciarra, S.: EU Commission green paper 'modernising labour law to meet the challenges of the 21st century.' Ind. Law J. **36**, 375 (2007)

Seifert, A.: Employee data protection in the transational company. In: Hendrickx, F., De Stefano, V. (eds.) Game Changers in Labour Law: Shaping the Future of Work. Bulletin of Comparative Labour Relations No. 100 (2018)

Solon, O.: Big brother isn't just watching: workplace surveillance can track your every move. The Guardian, 6 Nov 2017. https://www.theguardian.com/world/2017/nov/06/workplace-surveillance-big-brother-technology?CMP=share_btn_tw (2017)

Standing, G.: A Precariat Charter From Denizens to Citizens (2004)

Stone, K.V.W.: From Widgets to Digits: Employment Regulation for the Changing Workplace. Cambridge University Press, Cambridge (2004)

UNI Global Union: Global Union Sets New Rules for the Next Frontier of Work—Ethical AI and Employee Data Protection. UNI Global Union, 11 Dec 2017. https://uniglobalunion.org/news/global-union-sets-new-rules-next-frontier-work-ethical-ai-and-employee-data-protection (2017)

Zwolinski, M.: The pragmatic libertarian case for a basic income guarantee. Cato-Unbound, 4 Aug 2014. https://www.cato-unbound.org/2014/08/04/matt-zwolinski/pragmatic-libertarian-case-basic-income-guarantee (2014)

Roboadvice, Artificial Intelligence and Responsibility: The Regulatory Framework Between Present Scenarios and Future Perspectives

Pasquale Cuzzola

Abstract Artificial Intelligence offers peculiar development prospects in the field of electronic commerce, on-line banking and financial services: this is a typical event of FinTech, a phenomenon considered a driving force for progress and promoting the capital market union (primary goal of the European Union). In particular, the field of automation of financial advice has considerable relevance: the robo-advice and its "subspecies" ("pure", mixed, "robo for advisors") allow us to ask ourselves about the qualification of the existing relationship and that involves the investor, on the remedies available to him/her in the event of relationship disorders and liability profiles. This last aspect presents an element of complication due to the new "factors" introduced by the AI in the structure of the relationship, no longer simplified as direct and exclusive between two human beings. It is therefore necessary to investigate the possible development guidelines and to ask whether there is already, in the current legal system, an adequate regulatory framework or whether, alternatively, a direct intervention by the legislator is necessary. We could have: new interests on which to ask oneself about their legal recognition; or ways for a better pursuit of the already recognized interests thanks to new technologies; or again, the needs for regulatory intervention aimed to set up new cases of better protection of the same interests.

Keywords Roboadvice · FinTech · Liability in financial services · Automation in financial advice · Artificial Intelligence

1 Roboadvice in the Frame of the FinTech Phenomenon

The deepening of the roboadvice raises a series of interesting legal questions of identification and interpretation of the applicable regulatory framework; it makes the treatment, in the light of the current and above all the implications deriving from the evolution of artificial intelligence in this field, even more complex.

P. Cuzzola (✉)
Reggio Calabria, Italy
e-mail: pasquale.cuzzola@unirc.it

D. Marino and M. A. Monaca (eds.), *Economic and Policy Implications of Artificial Intelligence*, Studies in Systems, Decision and Control 288,
https://doi.org/10.1007/978-3-030-45340-4_8

The contribution that legal science offers in this field has a double declination: on the one hand, it can indicate to the operators of the sector a reference regulatory framework; on the other hand, it highlights research perspectives for more adequate forms of regulation,[1] in order to influence "the robotics market itself and ensuring congruent development with the values proper to regulation (Palmerini 2016).[2] In the background of the studies in this field lies the perception that "the development of robotics and artificial intelligence can lead to a distance so that much of the work currently done by human beings is done by robots" (European Parliament 2017). [3]

The roboadvice represents a typical expression of FinTech,[4] as a macro-implementation of different sectors (IOSCO 2017)[5] of the financial markets thanks to the help of technological devices that have not been described unanimously from a juridical and economic point of view yet[6]; for this reason a starting point is identified (more than a common definition), which frames it as "a new financial industry that applies technology to improve financial activities" (Schueffel 2016).[7]

Likewise, the origins of FinTech have not uniquely identified, although a link with the crisis of the last decade must be recognized with the consequent lack of confidence in the so-called "classic" financial system.[8] In any case, the FinTech has reached considerable dimensions globally so far.[9]

The institutions of the European Union themselves are interested in monitoring their development (European Commission 2016a) to identify benefits to be pursued and risks to be prevented and eventually to be addressed (Amidei 2017): the objective

---

[1]They are completely new, in terms of interpretative adaptation on the existing ones.

[2](P. 1818). The author refers in particular to the European context.

[3](P. 3 recital E). The document goes on to highlight how, nonetheless, there are also "concerns about the future of employment and the sustainability of social security systems if the current fiscal base is maintained, giving potentially rise to a growing inequality in the distribution of wealth and power".

[4]Word linked to the expression "Financial Technology".

[5]IOSCO (P. 4 Fig. 1) highlighted eight macro areas of operation of FinTech players: (1) payments; (2) insurance; (3) Financial planning; (4) equity and lending-based crowdfunding; (5) blockchain/DLT (cropt currencies, smart contracts, registration and asset tracking); (6) investments and trading (high frequency trading, roboadvisory); (7) research and analysis of information (Big Data, predictive analysis, etc.); (8) security (digital identity, encryption, fraud management).

[6]See, above all, the definition of the European Parliament (European Parliament 2017b): "FinTech should be understood as finance enabled or provided by new technologies, affect the whole financial sector, from banking to insurance, pension fund, investment advice and market infrastructures". One of the first attempts (in Italy) to deal systematically with issues relating to FinTech may be found in Paracampo (2017), p. 1 et seq.

[7]P. 46.

[8]This is mostly true with reference to Western markets, where in different contexts, such as those in Asian and African countries, FinTech has primarily represented an opportunity for economic development.

[9]This is demonstrated both by the diffusion and by the types of activities that are involved in it, as well as by people who go around to this macrophenomenon, thanks to which today we can talk about the "global fintech revolution": Kauffman and Ma (2015).

is to "develop a global action plan" (European Commission 2016b) for the exploitation of the opportunities of FinTech, linking this perspective to the purposes of the Union (such as the union of the capital market[10] and the digital single market,[11] with particular regard, in this case, to financial services).

In light of the difficult identification (and non-linearity) of the paths that the development of FinTech will follow in the future (even next), it could be useful, immediately, to worry not so much about imagining new operating rules but rather to thoroughly examine the current regulatory framework, in order to identify any deficient points but also possible hermeneutic link in possible future scenarios.

2 Financial Advice and Automation: The Contribution of AI

The specific phenomenon of roboadvice conveys the performance of a real financial service carried out thanks to the help of a data processing platform. These two indicators are not yet sufficient for its framing; more in detail, roboadvice can be defined as the financial consulting service offered in an automated way to the public through digital platforms based on algorithmic calculations.

Such a general definition allows us to include: both the more stringent meaning of roboadvice adopted by the European Mifid 2 Directive, according to which the service in question is "only that offered directly to the public […] on transactions concerning specific financial instruments or portfolios of financial instruments presented as suitable for a specific client"; and the broader one, accepted by the ESAs, including the services aimed at "recommending asset allocation of the financial portfolio for predefined types of investors and at providing general and specific information only for the execution of transactions through digital platforms; exclusively support the activity of the intermediary staff" (Linciano et al. 2019).[12]

The output provided by the robot—as a substitute or a support of the natural person intermediary, as we will see shortly[13]—can be considered a personalized recommendation when customers are guided "in an automatic online investment process through a set of filtering questions" (Linciano et al. 2019).[14]

The characterizing element of roboadvice consists in the possibility that the automated modality permeates the entire consulting process, from the genetic phase to

[10]It is an integral part of the third pillar of the Commission's investment plan for Europe.

[11]In this regard, the European Parliament has addressed an invitation to the European Supervisory Authority and to the European Commission, in the perspective of "achieving an efficient and competitive European financial system. That is more in-depth and more integrated, stable and sustainable": European Parliament 2017b, p. 7.

[12]P. 7.

[13]Both in case of being an independent consultant or a professional who works for a legal entity intermediary.

[14]P. 8.

the final one. It will be possible to manage with the help of the robot both the initial request for information to the potential investor, in order to outline the profile, as well as the final stage of elaboration of the investment recommendation, in addition to the intermediate phases and any subsequent phases of real investment.[15]

The most innovative profiles of roboadvice are therefore connected to the possibility of using digital platforms to manage the entire phase of the relationship and the provision of the service.[16]

The significant possibilities offered by online platforms in this sector (De Franceschi 2016)[17] are substantiated by their ability to neutralize the obstacles due to distances, thus favoring the convergence between demand and supply of subjects that once were unlikely to have been "encountered".[18]

At this point it is necessary to ask what kind of Artificial Intelligence is adequate for the implementation and development of the roboadvice.

The suggestion of a type of investment, basing on the innumerable variables to be considered on the profiles of the investors, involves a necessary processing of the data that cannot be based exclusively on indications and information made available to the robot when preparing the operating algorithm and introduction of initial information. It will be necessary for the robot to work through automatic learning, the c.d. machine learning, which allows it to learn both from the information collected for each time and from the accumulated experience over time.

It can also be reasoned in terms of supervised learning, to the extent that the robot is subjected to the "final results" intended as goals to be achieved: from the connections verified in the example processes indicated it will be possible to identify the existing relationships between input, output, and final result, so as to be able to elaborate—and consequently learn—an appropriate general operating rule.[19]

Likewise, a form of development of artificial intelligence underlying the operation of the robot could also be that of "reinforcement", in which there is no prior "education" with respect to the multiple types of output that can be processed based on the selected inputs. The propulsive thrust of increasing the skills of robot's intelligence, in this hypothesis, will not depend on the "received" education, but exclusively on the accumulated experiences (why not, on the errors).[20]

Introduced in these terms the reference context of the considered phenomenon, it is necessary to continue the investigation wondering what are the evolutionary

---

[15]If these are offered as an additional service compared to the pure consulting.

[16]Thus including not only the marketing of the service and the stipulation of the contract, but also the phase of the customer's requesting of information necessary for his profiling, up to the formulation of the investment advice.

[17]De Franceschi speaks about "innovation engines".

[18]Obviously, this does not only happen in the field of financial services; the potential for sharing also facilitate processes of knowledge of web surfers, as well as conscious purchasing choices in the broad sense. See in this regard (Colangelo and Zeno-Zencovich 2016; Sénéchal 2016).

[19]This will allow, whenever the robot receives a specific input, an increasingly suitable choice of output, offered to achieve the goal.

[20]An instructed robot with reinforcement learning will interact in a context with variable features; the paradigmatic example is the "scenario" of the financial markets.

perspectives, the possible goals, the interpretative and operational problems and, above all for legal scholars, if there is—and which is—the "regulatory coverage" and whether the current regulatory framework is adequate to regulate the phenomenon or if there is a need to revisit principles and systems of discipline.

For responding to this enucleated questions it is necessary to deepening of the two "components" of roboadvice: on the one hand, the financial advisory service as a typical case in financial market law; on the other hand, the peculiarities of the aforementioned service generated by the provision of the same thanks and with the aid of artificial intelligence.

The financial advisory service, in Italian law, is explicitly defined in the art. 1 paragraph 5 of the TUF (Consolidated Law on Finance)[21] as the "provision of personalized recommendations to a customer, at his request or on the initiative of the investment firm regarding one or more transactions relating to financial instruments". This framework proves to be a fairly faithful transposition of the European Mifid II Directive.[22]

The possibility of performing the service in question is reserved and subject to the issue of a specific authorization to perform services or investment activities pursuant to the Consolidated Finance Act, issued by CONSOB.[23]

This "*placet*" coming from the administrative authorities in charge implies the guarantee that the recommendation given by the consultant has as its object financial instruments[24]; it is personalized, as it is the result of a specific analysis of the investor's (personal, income-financial, educational) situation; it is presented as suitable for the recipient to whom it is addressed, as it is specifically tailored to its manifested characteristics, investment prospects and risks. These are the requirements that only a thorough knowledge of the financial markets and a solid corporate structure can ensure.[25]

A widespread recommendation for "advertising" purposes, as it lacks the aforementioned characteristics, is the subject of a free service, not subject to reserves and not bound by a series of obligations envisaged for financial advisors to protect investors.

Financial advice presupposes that the subject authorized to lend it has the necessary competences in the field of financial markets, with particular regard to the performance of the securities (of which to propose diversified operations); this presupposes the investment suggestion most suited to the client's profile.

The placement of financial instruments of the subject-intermediary for which the individual consultant-natural person operates is not a characteristic element of the

[21] Legislative Decree n. 58/1998.

[22] Directive 2014/65/EU. It entered into force on 3rd January 2018.

[23] CONSOB is the acronym of Commissione Nazionale per le Società e la Borsa, ie Italian national commission for listed companies and the stock exchange. See the art. 19 TUF which governs conditions, limits and procedures and that provides relative power to the Consob.

[24] The definition of the expression "financial instrument" is found in art. 1, paragraph 1, lett. t, TUF; its fundamental characteristic is the negotiability in the capital market.

[25] Also for a reassurance in terms of possible responsibilities.

consultancy. And it is for this reason that, more recently, the current legislation has been modified with the provision, for the intermediaries, of the obligation to specify to the customer if the advice given has or not an independent base.[26]

There is no written requirement for the financial advisory contract[27]; otherwise, any support, to the so-called "pure" consultancy, of the subsequent monitoring of the investment, requires that the relative agreement be formalized on a durable medium, not necessarily paper. [28]

The provision of consultancy can take place either independently or alongside another investment service.

In any case, both the appropriateness rule, also envisaged for all other financial services, and the adequacy rule apply to consulting.

The first is aimed at verifying the client's actual understanding of the complexity and risks related to the investment to be made, as well as his experience.[29] If the intermediary deems an investment that the client intends to pursue not appropriate for him, he has the obligation (only) to warn him, without having to proceed with other obligations.

Otherwise, the adequacy rule[30] requires a more meaningful verification and assistance activity. The intermediary will proceed to request more in-depth information from the client, including investment objectives, income and asset capacity, as well as to bear the costs of an investment and the possible losses arising therefrom.

In the absence of this information, or if the information framework suggests that an investment prospect is inadequate with respect to the client's profile, the intermediary is obliged to report it and to stop without proceeding.[31]

Since intermediaries are required to comply only with the rule of appropriateness and not adequacy for the provision of other financial services, the provisions on financial advice to which both rules apply, are more rigorous.

Once so shortly identified the advisory service in a synthetic manner, the profile of its automation can be taken into consideration.

---

[26]In the case of an independent base, the consultant has no ties with a producer, so his remuneration consists of a activity linked to the proveided consultancy; otherwise, the remuneration mechanism for retrocessions will operate.

[27]See Roppo (2009, p. 489), Perrone (2016 p. 204). Nevertheless, even in the light of secondary regulation, intermediaries are holders of information obligations towards customers who are almost always acquitted in writing: see Sfameni and Giannelli (2015, p. 90).

[28]The form requirement, in this case, is required for evidentiary purposes of the intermediary-investor agreement (see, on the subsequent recommendation, Article 58, paragraph 1, Delegated Regulation 565/2017, which limits the reference to the written form pursuant to Article 23 of the Consolidated Law on Finance, effective from 3 January 2018).

[29]In particular, the investor will be asked what kind of service and financial transaction he knows, or with which he has had experience, in addition to verifying both his level of knowledge and of the work activity he has carried out. A product may be definied appropriate if the customer demonstrates sufficient knowledge and experience to understand the risks involved.

[30]Planned for the portfolio management service, as well as investment consulting.

[31]See art. 25, paragraph 2, MiFID II and art. 54, paragraph 1, second period of the 2017/565 Delegated Regulation.

In this regard, a financial consultancy can actually be considered "robotized" when at the basis of data processing (before) and output (after) there is not a mere "calculator" but, precisely, a robot with artificial intelligence, which allows us not to limit ourselves to mere arithmetic calculations, but to collect and process data by imitating the human mind, thus proceeding with complex logical inferences and involving different types of data, information and skills.

Robotic automation, therefore, will occur if the device is able to perform its task without or with limited human intervention, if it uses the information collected by the investor to produce a recommendation independently and if the same recommendation has the characteristics of financial advice.[32]

A service provided by instruments lacking the aforementioned characteristics and capable of offering general information on the performance of the market, or which are substantiated in mere calculation tools, or in mere comparison, cannot be considered "financial advice".

The difficulty in identifying precise coordinates of identification of the automation of the output and distinguishing a simple calculator from a robot, is related to the same lack of uniqueness in the framing of the notion of robot at the scientific level (Uricchio 2019)[33]; however, it would be difficult to link the various robotic applications to a univocal and detailed definition at the same time.[34] "Despite these ambiguities, there is substantial agreement in identifying the distinctive characteristics of robots from a technical point of view in the ability to collect data through sensors (sense), to process raw data (think), to plan and carry out actions through the knowledge and

---

[32]According to ESAs (2015, pp. 12–14), the automation requirement occurs, in particular, if: "(1) The automated tool is used directly by the consumer, without (or with very limited) human intervention; (2) An algorithm uses information provided by the consumer to produce an output; (3) The output of the tool is, or is perceived to be, financial advice".

[33]For International Standards ISO 8373:2012, robot is an "actuated mechanism programmable in two or more axes with a degree of autonomy, moving within its environment, to perform intended tasks". The standard further specifies that a robot includes the control system and interface of the control system. The International Federation of Robotics (IFR) uses the same definition as the ISO standard.12 Nevejans suggests that a legal definition of robots be based on six conditions. In her view, a robot: (i) is a physical machine ("machine matérielle"); (ii) is alimented by energy; (iii) has a capacity to act in the real world; (iv) can analyse the environment; (v) can render decisions; and (vi) can learn".

[34]This constitutes a "common observation and constant starting point, even if negative, of every reflection on the subject" (Palmerini 2016, p. 1825). Some studies, above all those that have taken into consideration different categories of robots (for contexts of diffusion and utilization), have tried to frame them by classification categories. Among others, starting from the aspects more recurrent in the various definitions of robots present, the Robolaw project promoted by the European Union led to the identification of five categories on the basis of which evaluating and defining the robots: "The categories are as follows: (1) nature, which refers to the material in which the robot manifests itself; (2) autonomy, which refers to the level of independence from external human control; (3) task, which refers to the application or service provided by the robot; (4) operational environment, which refers to the contexts of use; and (5) human–robot interaction, which refers to the relationship established with human beings": Palmerini et al. (2016).

information acquired, as a rule based on pre-established objectives (act)" (Palmerini 2016).[35]

The peculiarity of artificial intelligence can perhaps be identified in the possibility of learning and it can be a driving force for its development but also a reason for the emergence of complex situations as the artificial intelligence itself evolves and gives rise to unusual scenarios and, therefore, difficult for operators to manage.

The provision of advice, in fact, is a classic activity that is refined and improved by learning from the experience and, in this way, acquiring ever greater competence. This process could well involve a robot.

3 Benefits and Risks

The propulsive impulses to the development of roboadvice are justified in light of the benefits that the same is hoped to generate: the possibility of accessing a service at lower prices than those ordinarily practiced by traditional advisors (Linciano et al. 2019)[36]; greater accessibility, bringing businesses and investors closer to the financial markets and expanding the range of services available; the streamlining of intermediation, which thus becomes "disintermediate".

It is imagined, indeed, that a limited (if not even absent) investor-consultant interaction, in addition to increasing the speed and convenience of the service (Hofacker 2001), contributes to reducing the psychological "pressing" on the client; furthermore, the digital context in which data is collected and processed would be able to reduce the information gap.[37]

Also the European supervisory authorities (ESAs),[38] based on the research conducted, have identified and indicated the benefits deriving from automation in financial advice (ESAs 2015).[39] The same institutions of the European Union have been working in virtue of the benefits glimpsed, aiming to promote the development of the sector, ensuring the reconciliation between the various interests involved (European Commission 2016).[40]

Nevertheless, the risks of implementing roboadvice, which impose a certain caution in the approach to the phenomenon, cannot be ignored.

First of all, it detects the risk of "too fast click decisions", which acts as a counterpart to the speed that was greeted positively just now: the same rapidity could

[35] P. 1825.

[36] P. 14.

[37] So called clickstream behavior: see Katawetawaraks and Wang (2011).

[38] It is made up of the European Banking Authority (EBA), the European Securities and Markets Authority (ESMA) and the European Insurance and Occupational Pensions Authority (EIOPA).

[39] The considerations were then resumed later: "the ESAs identified benefits of automated advice to consumers and financial institutions in respect of:—reduced costs for both consumers and financial institutions;—easy access to more products and wider client base for financial institutions; and—improved quality of the service provided" (ESAs 2018, p. 8).

[40] See p. 6.

generate an unjustified "frenzy" to proceed, thus eliminating the accuracy of the choice. Also, the absence of interaction could activate a boomerang effect for the customer, who instead of appreciating the context and feeling the psychological "pressing" reduced, could feel uncomfortable, consequently losing lucidity in his choices (Prasad and Aryasri 2009).

Still, it must not forget the dangers of information overload which is possible because of the digitized way of collecting and sharing information (Linciano et al. 2019).[41]

Finally, the dangers hidden behind an activity carried out by an unauthorized intermediary must be considered; they are less perceivable "at a distance" and with the "filter" of a robot,[42] or in the presence of conflicts of interest, especially where the algorithm has been already set in the programming phase to "direct" customers towards investment choices that are convenient for the intermediary.

More generally, the advent of a robotic element within the procedure aimed at financial advice involves a multiplier effect of the risks for the investor, which may not have clear features of the service, the nature of the information received, the identity of the subject providing the service: the technological dimension adds others to the "ordinary" ones in every investment decision.[43]

The risks associated with the roboadvice have also been identified by the ESAs in the "Report on automation in financial advice": "the ESAs identified risks of automated advice and financial institutions in respect of:—consumers with limited access to information and/or limited ability to process that information;—flaws in the functioning of the tool due to errors, hacking or manipulation of the algorithm;—legal disputes arising from two to unclear allocation of liability; and—the widespread use of automated tools" (ESAs 2016)[44]. The variability of probability and impact of their actual verification was also highlighted in the document referred to.[45]

In any case, a summary evaluation of the benefits and risks connected with the implementation of the roboadvice has to underline how they can be understood at best, in their concrete incidence, only if they are considered as two sides of the same coin. In fact, their specular presence is indicative: it can almost never be considered

[41] See p. 51, where a reference is made to the risks deriving from excessive reliance on the informal advice and from the c.d. framing effect; see also ESAs (2015), p. 21 et seq.

[42] The "disguised consulting" risks of eliminating the asset guarantee in the event of liability.

[43] Among the risks, the hermeneutical difficulties of allocating the costs of responsibility deriving from the increase in complexity of the case with the advent of robotic technology must also be considered; this aspect will be examined in detail shortly as a specific subject of this research.

[44] P. 9.

[45] There are peculiar outcomes of some researches which, indeed, on the basis of a statistical datum, show a series of critical profiles, potentially capable of severely limiting the development of roboadvice. In fact, it was detected the presence of some contractual "forms" with which it is noted that the device: does not provide a real financial consulting; it is not immune to conflicts of interest; it does not necessarily entail a benefit in terms of costs for the customer; does not act in the best interest of the latter; does not meet the safety standards required for fiduciary investments: see Fein (2015, p. 8). Furthermore, these "are not designed for ERISA retirement accounts and would not meet the DOL's proposed 'best interest' contract exemption".

that a benefit is present in a total and exclusive manner, without the possibility that the risk associated with it is emerging, and vice versa.

4 Different Models of Roboadvice and General Elements of Complexity

Both the positive and negative elements of roboadvice as well as the methods of manifestation of the service, can be investigated only taking into account some specificities.

Considering, in fact, the purpose of the service and the structure of the relationship as benchmarks, roboadvice manifests itself in a plurality of "sub-types", which contribute to increasing the complexity of the phenomenon generally considered and affect the methods of expression and effects (beneficial or unwanted) of the implementation of artificial intelligence in this area (as we will verify).

The "pure" model provides for the automation of all the phases of the service, from that of first contact with the investor, passing to the exchange of information (yields and collections), including the "profiling" phase, up to the true and formulation of the recommendation. The relationship exists between final investor and intermediary, qualifying as "B2C" (business to consumer).

The hybrid roboadvice sees the possibility of a variable combination between human interaction and robotic processing: it depends on the phase of the value chain in which the automated element is manifested, so that the manifestations of this type are also manifold. The common element is the qualification as "B2C" also for this hypothesis, being the service directed to the final investor.

The last model is the so-called "robot for advisors", in which the robotic activity represents an aid and a support for the sole consultant natural person who in turn uses it to carry out one or more phases of the financial advisory service,[46] thus maintaining the direct relationship with the end customer. Therefore, the latter does not come into contact with the robot. The robot for advisor type, therefore, exists between professionals ("B2B").

To date, the pure model is not widespread, whereas the combination of the results of artificial intelligence with human intervention represents the prevailing hypothesis.[47]

---

[46]Collection of information on the customer, processing of the same for the purposes of assessing adequacy, asset allocation, portfolio selection, up to the formulation of the "final" recommendation and, possibly, negotiation and portfolio rebalancing, if envisaged as additional activities with respect to the mere advice.

[47]In this sense a summary of the current situation can be found in ESAs (2016, p. 17): "The majority of respondents to the DP that are currently offering financial advice actually use the hybrid business models. In particular in the securities, respondents indicated that the main two types of tools:

• "Fully automated tools": these are completely automated tools, driven by algorithms or decision trees, with no or very limited human interaction involved in the advisory process. In response to the information provided by the consumer, the recommendations are based on the following criteria:

In all the various possible models, some elements of specialty that characterize the phenomenon can be identified increasing the degree of applicative and interpretative complexity, and of which it will be necessary to take into account in order to proceed with a verification of the applicable standards and the remedies available, especially in case of side effects.

On the contrary, the assessment of compatibility of the regulatory framework in force with regard to automated consultancy precisely needs to take it into account as a priority.

The first peculiar aspect is the structural increase of the subjects involved in the financial consultancy in the case of its provision (in whole or in part) through automation.

Depending on the type of roboadvice model, the nature of the subjects involved may vary, but their quantitative increase remains constant. In fact, not only the customer and the intermediary/consultant are in relief, but also the owner of the robot/software (if different from the intermediary), or the supplier (for other reasons), or the developer of the algorithm on the basis of which artificial intelligence works.[48]

The overall process of automated consultancy ultimately sees a multiplication of actors.

A further element of complexity is naturally represented by the evaluation of the incidence and impact that Artificial Intelligence will have with the progress of developments in this field.

The growing and increasingly rapid evolution of AI makes it difficult to anticipate and abstract the ability of robots to assess appropriately the situation of investors, but also to rebalance their profile on the basis of the data collected, perhaps noticing some defect and requesting further information.[49]

Finally, the specificity of roboadvice is found in the novelties that can be verified in point of fact causative of any undesired effects. In the absence of automation elements, the exclusive presence of a natural person and not a robot intermediary presupposes

---

• "Hybrid advisory tools": These tools combine an automated tool (algorithm or decision tree) with the ability to interact with a human advisor. Certain business models always involve a human advisor, typically to provide additional customer service, and/or to provide an additional quality overlay to the recommendations presented by the automated tool. Respondents to the DP also noted that, in some business models, a human advisor would be engaged when it is considered that the consumer has more complex needs than those that the automated tool can accommodate.

[48]Formally, the "subjects" involved, from a purely legal point of view, could also remain just two, for example in the event that the consulting service is provided by a brokerage company that develops the software for the operation of the roboadvice by it own, perhaps with internal staff. Nevertheless, the necessary involvement of people other than those who are the protagonists of the "classic" financial advisory relationship remains firm, though in phases before (or after) the ordinary "value chain" of the service.

[49]While in traditional advice the client's profile can be defined and then rebalanced on the basis of further questions to meet all individual needs or to clarify any inconsistencies, an automatic process based on a set of standard questions could have a lower capacity of getting the peculiarities of the individual and incorporate them into the proposal.

that at the basis of adverse events only human behavior (activities connected to the consulting service, in any of the phases, provided inadequately) can be found.

Except that, since artificial intelligence allows the exercise of activity without any human intervention, in the case of roboadvice any unfortunate consequences in the provision of the service, even if identical in terms of interests worthy of legal protection would see a different originator of the injury.

This will not be (more), that is, "simply" a negligent, imperfect, reckless, or possibly preconceived human act[50] in favor of favoring a particular subject, in the case of non-independent consultancy (perhaps violating the law on the conflict of interest).

On the other hand, a malfunction of the algorithm may occur when the data is processed, resulting in an unsuitable recommendation with respect to the client's profile.[51]

Complexity factors highlight both the potentialities and the problems that have to be faced with regard to the effects of the advent, development and evolution of artificial intelligence in this sector.

Moreover, the risks to be faced may manifest a relationship of inverse proportionality between frequency of verification and difficulty of solution. And in fact, the more the artificial intelligence applied to roboadvice evolves, the more the risks of undesired effects and damages (especially to potential investors) decrease in quantity, but increase in degree of complexity, becoming increasingly difficult to face and solve, as generators of doubtful cases of not easy solution.

5 Assessment of Compatibility of the Current Regulatory Framework. Focus on the Remedies for the Investor

The verification of possible solutions in doubtful cases requires a prior analysis of the current regulatory framework (at national and supranational level) in the field of financial advice, to test the estate when it is put "to the test of roboadvice".

A primary regulatory reference is the European Mifid I and II Directives.[52]

Already the art. 19 of the first prescribes investment firms to: act honestly; provide information, clear, fair and not misleading, therefore appropriate and understandable; communicate appropriate guidelines and warnings on the risks associated with the investments. To these latter obligations of communication directed by the intermediary to the investor, there are added those in the opposite direction: the intermediary will have to obtain information relating to knowledge and experience of the investor

[50]Evaluation and/or subsequent declaration, understood as the final manifestation of the consultancy.

[51]Indeed, IOSCO (2017 p. 33), gives note how, in addition to the damage profiles, in any case differently programmed algorithms generate diversified outputs even starting from the same investor profile.

[52]Respectively: Directive 2004/39/EC; Directive 2014/65/EU. The Mifid I was implemented by Italian law with Legislative Decree 17 September 2007, n. 164.

about investments, as a necessary precondition for formulating a recommendation appropriate to his profile.

It is possible to inform customers in a standardized way.[53]

Concerning the recent Mifid II directive it is necessary to remark[54] in particular the art. 24, par. 2, which provides for the obligation[55] for companies to offer and recommend financial instruments "only when this is in the customer's interest".

The aforementioned legislative references, given their technological neutrality, show full compatibility with "new" methods of providing the financial advisory service, thus making peaceful the implementation of roboadvice in this sector, especially according to the most common models (hybrid and "for advisor").[56]

The same conclusion is reached by restricting the focus to the internal legislation of the Italian legal system.

The most automated models of roboadvice (pure and, to some extent, hybrid) turn out to be financial advisory services rooted in an e-commerce contract (Bravo 2012),[57] to which the discipline of distance marketing of financial services applies (articles 67-bis and ss. cons. code), favoring the investor, where he is a consumer.[58]

Beyond the general provisions on the services contained in the aforementioned regulatory apparatus, the compatibility check of the roboadvice passes through the analysis of the applicable rules such as remedies available to the investor in the hypotheses of "pathology" of the relationship. These aspects do not deal exhaustively with the regulation in question, but may refer to the general discipline.

A hypothesis of nullity of the contract is foreseen by the art. 67, paragraph 4 of the Consumer Code, in the event that the professional "violates the obligations of pre-contractual information so as to significantly alter the representation of his characteristics".

This form of invalidy is relative nullity in question is relative: only the consumer can make it worth.[59] His specialty derives from the fact that, usually, so far as the intermediary violates the information obligations, the contract would not be invalid but only the remedies of responsibility and compensation would be activated.[60]

---

[53]See in this regard, in addition to art. 19, paragraph 5, also art. 21, paragraph 4, as well as art. 44, paragraph 1.

[54]Beyond some news that does not put here any emphasis on highlighting.

[55]Although it is formulated in a peculiar manner, given that it is expected, literally, that intermediaries should "ensure that".

[56]See Linciano et al. (2019, p. 81), according to which this would not require an immediate rethinking of the regulations. For similar assessments in comparative perspective, see Hunt (2017).

[57]See pp. 255 and 260.

[58]Article 3, paragraph 1, lett. a, of the "Consumer Code" (Legislative Decree 6 September 2005, n. 206) defines the consumer as the natural person who acts for purposes unrelated to the business activity (or craft, or professional) eventually carried out.

[59]There is also the duty for people to return the property they got, without prejudice to the further damages granted to the consumer.

[60]Two sentences of the Supreme Court with joint sections (Court of Cassation 2017 n. 26724; Court of Cassation 2017 n. 26725) have emphasized that, in the field of financial services coomercialized

Outside the special hypothesis just mentioned, the violation of the obligations imposed on the intermediary, even if committed through automated means, will lead to the pre-contractual or contractual responsibility based on the time of verification of the violation (in the training phase or in course of contractual execution[61]), with consequent obligation to pay damages and possibly terminate the contract.

The verification of the existence (before) and the quantification (then) of the damage suffered by the investor will be carried out according to the traditional canons: whoever complains about the damage will have to invoke compensation indicating that he has suffered a decisive violation for the conclusion of an operation then it turned out to be ruinous, but which it would not have run had it been properly informed. The burden of proof will be distributed according to the provisions of art. 23, paragraph 6, TUF[62]: thus, the investor will be able "attach the non-fulfillment of the aforementioned obligations by the intermediary" and provide proof (only) "of the damage and the causal connection between this and the breach, also on the basis of presumptions"; the intermediary, in turn, will have to prove "the fulfillment of the specific obligations imposed on him, attached as unfulfilled by the counterparty and, from a subjective point of view, to have acted with the specific due diligence required".[63]

On the evidential level, where the advice is given to a consumer, the latter will be recognized as having additional advantages: among these, consider the provisions of art. 67-vicies semel c. cons.,[64] which identifies the duty of the consulting service provider to demonstrate the fulfillment of the information obligations, the performance of the consumer's consent to the conclusion of the contract and the execution of the same; a presumption of harassment is also envisaged for clauses that derogate from the provisions on the distribution of the burden of proof.

Compensation for the damage will be accompanied by the termination of the contract, such as the discharge of the negotiation constraint due to an illness that has occurred,[65] operating in the corresponding performance contracts, when the necessary information has not been provided, or the inadequacy of a financial transaction, in violation of the provisions of art. 21 TUF: the resolution will operate mostly when the advice is given on an ongoing basis, more rarely when it is "isolated".

---

at a distance, it is not possible to identify a general rule that penalties the violation of obligations of pre-contractual information through the use of the sanction.

[61] For more information on the nature of liability, see Maggiolo (2014), p. 497.

[62] According to which "in the compensation claims for damages caused to the client in the performance of investment and ancillary services, the burden of proof of having acted with the specific due diligence required is up to the authorized parties".

[63] The expressions are used by Court of Cassation 2016, n. 810, which conforms to an orientation in this sense expressed by the Supreme Court (see, among others, Court of Cassation 2009 n. 3733; Court of Cassation 2015, n. 826).

[64] Legislative Decree 6 September 2005, n. 206.

[65] As a remedy operating both in the event of termination and invalidity of the contract, pending the existence of an illegal act performed by the investment service provider: see Maggiolo (2014 p. 508), according to which compensation is defined as "exclusive or concurrent protection … always available for the customer damaged by the pathological behavior of the intermediary".

It is also theoretically applicable the remediation of the annulment of the contract,[66] as a less radical type of invalidity and deriving from a defect of consent in the process of forming the will of the parties, in particular the investor. This defect could also occur where the provision of the service takes place in the form of roboadvice: in fact, while the occurrence of moral violence is substantially impossible, otherwise, situations of error may arise—spontaneous or generated by fraudulent (and therefore malicious) acts—generated by the appeal artificial intelligence underlying the robot.[67]

Indeed, any failure to understand the peculiarities of the service offered, as well as the automated aspects of its performance, risks being amplified precisely because of the same automated methods.

Compensation for damages[68] could be reduced, or even excluded, in the event of negligent behavior by the same investor who, alone or in competition with other contributing causes, caused the damage: note in this regard the art. 1227 cc,[69] a rule that has to be applied with particular prudence, since "the special brokerage relationship necessarily implies a degree of reliance on the professionalism of the intermediary and, therefore, on the adequacy of the information provided by him, which would be contradictory to balance with the burden of the same customer to directly take information from another source" (Court of Cassation 2016, n. 8394).

The implementation of the roboadvice is compatible with the regulatory framework outlined. None of the provisions referred to manifest problems of applicability in the case of automated advice, where the activities of the intermediary are carried out, in whole or in part, by a device equipped with artificial intelligence. Possible remedies and indemnifiable damage remain the same, rather requiring only an interpretative and applicative adaptation, due to the peculiarities of the roboadvice deriving from the elements of complexity highlighted above: the increase of the subjects involved, the impact of artificial intelligence, the different originator of the injury.

The latter, however, do not lead to a change of perspective on liability allocation.

---

[66] See Court of Cassation 2009, n. 3773.

[67] It is a case of annulment (not nullity) ruled by Court of Cassation 2007, n. 26172; (Ticozzi 2007; see already also Scalisi 1994, p. 190).

[68] Indeed, if the error is caused by the violation of the information obligations in the pre-contractual phase by the advisor and if the investor is a consumer, he would find the most radical hypothesis of nullity of the contract, by virtue of the art. 67-septiesdecies, paragraph 4, c. cons.: but only if from this violation derives a significant alteration of the representation of the features of the offered product.

[69] The mentioned norm is entitled "Competition for the negligent event of the creditor" and provides, in the first paragraph, that "If the negligent fact of the creditor has contributed to the damage, the compensation is reduced according to the gravity of the fault and the extent of the consequences that derived from it." On its abstract applicability see Court of Cassation 2006, n. 8229.

6 Responsibility and Relationships Between the Investor-Intermediary-Software Provider

In the event of pathological outcomes of the advice which see the unfortunate investor damaged, the holder of the advisory legal relation will be in any case the liable, acting as intermediary, regardless of the circumstance for which the robot that has (in all or in part) provided the activities necessary for the provision of the service have been developed in-house or acquired by third parties.[70]

This is what emerges in the legal system in light of the current situation of evolution of the phenomenon of automation of advice, as well as the need to offer guarantees of certainty to the investor on the possible remedies and on the "counterpart" to turn to.

Even European legislation would seem to be oriented in this direction: the art. 31 of the EU Delegated Regulation n. 565/2017 states that "investment firms that outsource essential or important operating functions remain fully responsible for compliance with all the obligations imposed on them by Directive 2014/65/EU".

Nevertheless, this conclusion does not prevent the intermediary, if it is a different subject from those who have supplied (for various reasons) the software with artificial intelligence, to claim the liability for having supplied a robot (and an algorithm for its operation) later proved to be "defective" and in any case a cause of damage.

Thus, it would be necessary to sue to obtain relief for the damages suffered, consisting of the costs incurred to compensate the investor but also in any further prejudice to those connected.

Based on the relationship with the provider of the algorithm,[71] the actions that can and must be undertaken by the intermediary will change (Linciano et al. 2019).[72]

The latter could have purchased the software for operating the robot equipped with artificial intelligence. In this case, in the Italian legal system, the rules regarding the defects of the product sold,[73] formally applicable, do not guarantee protection to the intermediary considered responsible towards the investor for the damage caused by the purchased product. But in some cases the Supreme Court of Cassation[74] has come to consider the existence of the "concurrence of a compensatory liability "*ex contractu*" with an "Aquilian" type [of the seller] if the damage claimed by the buyer pursuant to art. 2043 c.c. is configured [...] as a further consequence of the

[70]In this regard, it was stated that "the human behind the robots to be held responsible for its actions. The choice of the owner/user and the producer" (Bertolini 2013, p. 227); and again (p. 230): "furthermore, there may be conditions where it could prove useful to hold the owner or user because it is one of the best position possible to intervene and avoid harm, or rather compensate the damage when it occurred, irrespective of whether he was at fault in causing it".

[71]Or in any case of the whole or part of what is necessary for the provision of automated advice.

[72]P. 87.

[73]The warranty for defects (Article 1492 of the Civil Code) would not satisfactorily include the abstractly verifiable situations; moreover, the forfeiture and prescription terms (art. 1495 c.c.) would be too short, and mostly all the limitations to compensation for damages deriving from the defects of the sold product would be excessive (art. 1494 c.c.).

[74]We refer to Court of Cassation 2008, n. 11410; Court of Cassation 2014, n. 3021.

malfunctioning of the "res", which in turn produces autonomous and independent harmful events" (Court of Cassation 2005, n. 8981).[75]

Another possibility for the intermediary is the use of outsourcing[76] of the entire service or of well-defined phases of the same, thanks to the professional services offered by specialized subjects that allow the decentralization of the activity (Macario and Addante 2014).

The rules of the contract that are closest to the parties' forecasts, including the work contract, subcontracting, or contract, will apply to the report; they will also affect the rules of responsibility by virtue of which the outsourcer may be held responsible for the damage caused by the part of the outsourced activity and, to be precise, attributable to him. However, it is not easy to prove such circumstances, given the widespread practice of specifying that obligations are means and not results, with consequent limitation of liability.

Finally, there may be a user license agreement between the intermediary and the software provider, by virtue of which the former uses the immaterial device (equipped with artificial intelligence) but the latter remains the owner, receiving a sum as compensation.

The software could also be specially "cropped" on the needs highlighted by the user.[77]

The contractual relations just illustrated can also be used in the hypotheses of roboadvice that involve different subjects—with the aforementioned "multiplier" effect of the same—in the value chain. The search for possible responsibilities among the subjective profiles referred to would thus see possible novelties in the "starting point" ascertained as harmful, while the criterion for attributing it could then follow the traditionally operating rules. The real challenge of innovation lies precisely in identifying the new originators of damage and the criteria that can serve as an aid to the jurist for their legal qualification.

Indeed, the definitive placement of responsibilities by the final supplier may not always appear to be a sign of justice and fly for the development of the phenomenon. Nevertheless, the risk of a simplistic propensity to do so in the regulatory systems was reported. The "Report on automation in financial advice" has highlighted the

[75]More recently it was expressed in the same way Court of Cassation 2017, n. 16654.

[76]The reference to the outsourcing is understood as a variegated category of contracts, diffusing over time and becoming socially typical even before of finding a legal formalization, though they need to protection. They are onerous and for corresponding services, subjected to one or more types of contract (with the possibility of a mixed or complex contract) based on the concrete agreements of the parties. This includes the various possibilities for the company of rejecting the direct management of sections of the activity not included into the core business. For example, the business owner will be able to choose whether outsourcing the fulfillment of the service, or selling a branch of the company: see Court of Cassation 2006, n. 21287.

[77]For this reason, the relationship can be better defined as a "software development contract". Studies in this regard date back to several decades ago: see above all Rossello 1984. It is not always easy to identify the differences between the software development contract and outsourcing. In the first case, the intermediary has an "instrument" (even if it is an algorithm) with which he produces immaterial activities constituting the service he is providing; in the second case, it is the activity itself (in whole or in part) that is outsourced, therefore it is carried out by others.

following: "on the basis of legal assumptions, two to one allocation of liability, they were based on the assumption that it may be more likely that different financial institutions may be responsible for different parts of the service offered to online consumers in a disintermediated environment. Where there are no specific legal agreements on the ongoing basis, and if there are not appropriate controls in place over any outsourcing arrangements, it could happen that the financial institutions inappropriately delegated their regulatory responsibilities to the end provider" (ESAs 2016).[78]

Perhaps also because of the uncertainties connected to the risks above mentioned, there were also decidedly negative opinions on the current regulatory framework, despite having found a tendential adequacy of the same, even when worthy of some adaptations. Although not with specific reference to the phenomenon of automated advice, but looking at the world of robots in general, the draft report with recommendations to the Commission on civil law rules on robotics believes that there are "obvious [...] shortcomings of the current legal framework in the area of contractual liability, in sofar as machines designed to choose their counterparts, negotiate contractual terms, conclude contracts and decide whether and how to implement them make the traditional rules inapplicable, which highlights the need for new, more up-to-date ones" (European Parliament 2016).[79] Such an approach inevitably generates a lot of uncertainties among practical operators (even before that among jurists), with a consequent risk of curbing the evolution of the phenomenon.

7 Current Regulatory System and Potential

In reality, if it were not for the advances that artificial intelligence can make in this sector, considering, in a static perspective, the only current state of automated consultancy, the current regulations would perhaps be sufficient to guarantee coverage, legal certainty and protection in the abstractly verifiable situations (Decker 2017).[80] The current regulatory framework (both internal and supranational), although it is not very recent, would preserve a tendential adequacy, also because of its technological

[78]Pp. 10–11. Also relevant is the subsequent observation, which reports the results of the survey conducted on practical cases: "Respondents noted that disputes—related costs and future liability could be a barrier to automated advice for some institutions. Some respondents expressed responsibility over the allocation of liability when consumers receive advice via an automated tool, particularly, whether the liability can be attributed to consumers, providers or third parties. The outsourcing of functions to special providers such as entities in the FinTech industry was also a response to this risk".

[79]Recital X.

[80]P. 155. The author, referring to the European project "Robolaw", underlined the negative answer to the question: "do we need a special regulation for robots? [...] the conclusion is that no particular law for robots is needed. The technical performance of robots has reached the level of "ethical agents".

neutrality[81] that would allow the identification of an applicable discipline "regardless of the channel used to implement it, determining a tendential indifference to the traditional or Fintech method, with respect to the identification of the applicable remedies in case of violation."[82]

In any case, it would be necessary to avoid an "exceptionalist approach, […] typical of those who consider a priori the current norms inadequate to regulate the issues that emerge from technological developments, believing, therefore, always necessary to create new ad hoc regulations" (Santosuosso et al. 2012).[83]

Nevertheless, these conclusions manifest all their fragility where they are put forward in a future and evolutionary perspective. "Quid iuris" if the robot for the automated advice, thanks to the ever increasing (and faster) development of artificial intelligence, would learn more and more to "operate" in this specific sector, bringing to the extreme consequences the potential of full automation of the service of investment advice?

The regulatory cases referred to up to this point, would begin to show all their limitations, making them fail, or at least revealing the inadequacy of norms which, although they are general and abstract, could not have foreseen what the human reality was not yet capable of imagining, or at least (again) to configure as real and possible. The elasticity pursued by the principle of technological neutrality would irreversibly fail.

At that point, one wonders how the hypotheses of malfunction or damage would be managed and regulated in the presence of "expert" robots.[84]

Technological evolution leads us to consider the scenarios envisaged not so far away, as it is now possible that "in the space of a few decades artificial intelligence will exceed the human intellectual capacity" (European Parliament 2016a).[85]

However, the doubts raised—and which will be explored shortly—do not eliminate the awareness that the gradual weakening of the eminently human contribution, with the growth of the "computer-intelligent" component (Linciano et al. 2019),[86] could also make the operations performed not only faster, but also more reliable

[81] European Parliament (2017a), point 6, expressed itself in favor of the principle of technology neutrality also for the setting future regulation of the subject.

[82] Thus, where it's adfirmed a link between the technological neutrality and the absence of "significant deviations" of the automation financial advisory service from the more general phenomenon of the advice: see Linciano et al. (2019, p. 69).

[83] P. 497.

[84] We will return soon to a similar qualification. Also European Parliament (2017a), point 29, highlights how a massive use of automation in data collection and processing through algorithms amplifies the difficulties in identifying (and subsequently managing) the hypotheses of responsibility in the roboadvice.

[85] European Parliament (2017b, pp. 4–5). This point is confirmed even if we don't agree with the perspective that it will come "the point that, if we are not prepared, it could be dangerous for humans' ability to control what they have created and, consequently, also to control their capacity to be responsible of their own destiny in order to guarantee the survival of the species".

[86] P. 81.

and—hopefully—safe (Bouyon 2017). The hope is that, among the possible configurations of innovative processes, the main goal of technology is always to serve the humanity.

But some stages are still missing before reaching the prospective planned, which are not yet guaranteed by roboadvice but which the development of artificial intelligence could allow to reach.

To this end, a correct "education" is needed, to be understood as an adequate set up of the algorithm, as well as constant progress thanks to the experience gained in the automatic processes: "the use of robo-advisors requires training and education" (Fein 2016),[87] as indeed happens for a natural person consultant, whose experience and competence will be achieved over time and with "practice".

Artificial intelligence is the key to giving the robot autonomy in operations and the ability to learn on its own; these characteristics are considered "essential turning points for the assessment of liability" (Bertolini 2013).[88]

Nevertheless, for the purpose of resolving doubtful cases in terms of responsibility, it would be necessary to distinguish a strong and a weak notion of autonomy (Uricchio 2019): "the former, derived from the philosophical notion of moral agency [...], would surely force a change in the existing legal paradigm; the latter, the technical aspect of the control system of the robot [...], is not sufficient to justify a change in the law, and the comparison made with animals appears to be misleading [...]" (Bertolini 2013).[89]

Likewise, the robot's ability to learn could not, by itself, lead to a change in responsibility patterns.

Consequently, the transition from the progress of the robot's activities, favored by artificial intelligence, to the change of rules and criteria for the attribution of responsibility in doubtful cases will not be automatic.

8 What Criteria for the Imputation of Liability? A Proposal

All the raised issues can be addressed only by keeping in mind the coordinates of value and the criteria of liability imputation consistent with the legal system and consolidated culture, according to what the interpretative evolution has highlighted in this sector.

The reference principles for the choices are: priority investor protection as a "weak" part of the financial advisory relationship; the good performance of the

[87] P. 7. To date, in fact, "Only Financial Professionals Can Provide Portfolio Analysis in light of all the relevant factors a fiduciary must consider [...]. While digital tools can assist financial professionals in determining clients' profiles, the report indicates that prudent investment advice requires human judgment beyond such tools (although human judgment still depends on the skill of the professional).

[88] P. 217.

[89] *Ibidem*.

market and the non-distortion of its operation; attention to avoiding conflicts of interest and punishing behavior that does not respect the "rules of the game".

Principles must be followed by appropriate models of imputation of responsibility, as guidelines—in the event of unjustified damage to the interests of unfortunate investors—of the choice of the subject on which to allocate the cost of their restoration.

The system must be able to identify who is able to better bear the relative cost, according to an orientation that reflects the need to derive the consequences of one's own actions, omissions, but also entrepreneurial choices and investments (not necessarily guilty), without eliminating the need to promote a sector and a phenomenon whose benefits have been recognized for the development of society and without the costs in question being left "falling" on the weakest categories of the value chain, where this does not correspond to shared values of justice and fairness.

A double criterion or "binomial" is proposed for the allocation of the costs of the damages, that is for the identification of the person responsible, in the head of the person most suitable for their tolerance.

The choices in terms of responsibility can be oriented by identifying who, in a given action, has (or has had) the opportunity to act and enjoys an undoubted (perhaps the greatest) benefit deriving from the use of the robot.

The first criterion about the possibility of acting, must be identified not as the "duty" to take action, or to avoid a certain consequence, but as an objective exercise of a specific action incident on the value chain and by virtue of a de facto power, legally founded. Those who intervene in the programming or education phase of the robot, in the algorithm configuration, in the sale of the same, in various uses can act. The challenge lies in the search for those who, among others and more than others, can exercise an activity aimed at control and, therefore, to avoid unwanted episodes of damage.

The technological environment certainly makes this verification complicated. It is not easy, in fact, to verify whether, in general or even in a single episode, there is or not a dose of unpredictability of the path that could follow or that the artificial intelligence has followed and if a possible moment of rupture is due to the phase of original programming of the algorithm, or of "education" (perhaps subsequent to the sale) of the intelligent robot, or of the one in which the robot autonomously learns, or if it is completely independent as it can be traced back to determinants and non-avoidable external factors.[90]

But this verification is a duty for the interpreter; the complexity underlying it does not presuppose a structural impossibility.

Due to the difficulties just highlighted, the need arises to identify another criterion, which represents a moment of reinforcement but also a turning point in situations of "impasse" in the reconstruction of the case.

[90]Decker (2017, p. 155), referring to some researches conducted by Susanne Beck, highlights how "adaptive learning robots are thought to interact with humans in normal environments, and, consequently, they can react to new sensory inputs in an unpredictable way. If that is the case, on can hardly assume that this reaction was caused by a wrongful act of the programmer, producer, or even the user".

The search for an optimal allocation of damage costs, then, can be conducted through the criterion of the benefit deriving from the use of the robot. At the end of the benefit there are many situations of enjoyment, whether direct or indirect, whether patrimonial or not, associated with the artificial intelligence that makes the robot work.

Who enjoys it: those who create the algorithm and therefore sell their skills; who makes financial investments, who is the owner and sells the finished intangible product; who organizes to provide an outsourced software service; the intermediary that uses it to support its advice (saving time and ensuring greater reliability of the service provided); the intermediary who uses it to automate (almost completely) the provision of the service; the final investor for a (potentially) more facilitated use of the financial instruments market, and perhaps also for cost savings.

The newly resigned enjoyment situations diverge in terms of assumptions, costs and effects; but they are united by the existence in themselves of the benefit deriving to the subjects of the use of the robot equipped with artificial intelligence. It is understandable why enjoyment does not consist only of an increase in assets and why it can also be indirect.

The correlation between enjoyment (direct or indirect) and responsibility is linked to the idea—typical of common feeling, even before it is legally formalized—so that a person who has a source of benefit will normally also be required to take into account any costs from that deriving, according to the well-known adage according to which cuius commoda, eius et incommoda.

The search for the optimal solution for the allocation of costs passes, then, through a contextual verification of the reciprocal influence of the two proposed criteria, arriving at a sort of operation (almost algebraic) that has to be conducted having regard to the entire value chain in the automated advice: the bet is that the subject to whom the "greater algebraic sum" of the criteria indicated can be identified, in order to identify, therefore, in this way, also the most suitable criterion and subject for the attribution of responsibility.

The rules on the subject, if well defined at general level and well applied by practical operators, make it possible to "allocate the costs", with possible "displacements" (horizontally or vertically) that allow the subjects involved to be able to better internalize them. The final effect should be to implement a social and generalized well-being, at a triple level: wealth distribution; security implementation and risk reduction; insurance for catering to the unfortunate and blameless damaged (the investments conducted by those who know they can be called to responsibility will move in this direction). "By shifting the cost connected with an adverse event, liability rules and the wrongdoing to internalize the consequences of his actions and choices. Theoretically the adoption of the correct liability rule should ex ante induce the socially desiderable behavior, in terms of reduction of number of accidents and increase in safety investments, and ex post ensure recovery of the harm suffered by the individual through the action of the wrongdoer".[91]

---

[91] Palmerini et al. (2016). This is evidenced according to the perspective of a necessary organization of the regulatory framework "in order to balance opposing interests, but also—once desired policies

We are aware—and it is better to point out—that the practical implementation of the imputation criteria, on the one hand could be substantiated in a policy choice for the legislator; on the other hand, it imposes on the law a necessary aid and contribution from the other branches of knowledge, those that today are dealing with both promoting the evolution of artificial intelligence and finding the most hidden algorithmic mechanisms of functioning.

A renewed juridical order in this field would allow to face in a more systematic way a series of possibly verifiable hypotheses, with which the legal science as soon as possible will necessarily have to interface.

9 Towards the Future. Interpretative and *de iure condendo* Solutions

Coming to the point of the present research, addressing—without pretensions to completeness but with a paradigmatic approach—some issues that arise, or could soon be presented, to the interpreter.

First, we will examine a couple of hypotheses to be considered in perspective de iure condendo. Later, there will be a couple of possible scenarios that are difficult to solve, generated by the inexorable evolution of artificial intelligence.

The first question on the future development of the law concerns an explicit recognition of forms of strict liability in the cases of automation in financial advice.

The advice itself, as a financial service, is a negotiating relationship with the consequent genetic applicability of the rules on the contracts between intermediary and investor.

Nevertheless, it's possible a reasoning on the applicability of an objective extra-contractual liability starting from the hypotheses in which the provider of the robot with AI is a different person than the one who formally provides the advisory service: in this case, one could almost imagine a "double track" of liability imputation, with recognition of the provider's strict liability, which with the investor has not concluded any type of contract, in the event of errors and/or malfunctions of the robot attributable to the software programming phase.

A legal experience of reference could be that of liability for a defective product, with appropriate adaptations.[92]

---

are identified—taking into account the concrete effects and impacts of the rules on the market, not entirely related to general assumptions and unverified considerations about their presumed—or expected—consequences".

[92] […] If they are produced, the liability from defective product should be revised, otherwise it is not adequate to protect the interests involved. But the project of European standards on robotics (recital U) has already proposed possible itineraries in this direction: "manufacturers, owners or users could be considered objectively responsible for the acts or omissions of a robot if, for example, the robot has been classified as a dangerous object or within the area of product liability rules".

In this regard, the question remains whether what we have just proposed is already possible in the light of current law, or whether it is only conceivable "*de iure condendo*". The situations considered above would seem to orientate towards the second hypothesis: the injured investor can only contact the intermediary with whom he had concluded the advisory legal agreement. It is the same intermediary that, to date, has seen the possibility of activating the remedy of non-contractual liability towards the "supplier", even though in the particular form of compensation for damage to the image.

The new element would be the recognition of the responsibility of the robot provider with artificial intelligence, with direct legitimacy to sue recognized to the investor; it would be a further remedy available to the latter.

Thereupon, it would also be necessary to reason in terms of "*beneficium excussionis*", in order to avoid over-protection: it does not seem linear, in fact, that the investor can indifferently turn to one (intermediary) or the other (software provider) without a regulation of the various possible hypotheses and subsequent recours mechanisms. A regulatory choice in this sense could be facilitated by the two criteria proposed here.

In any case, an approach of favor for the recognition of strict liability is advocated by the Project report of European standards on robotics (European Parliament 2017b),[93] although in general and not with specific regard to roboadvice.

The most delicate question for the scholars and which has always influenced research on robotics in general is about the possibility, the opportunity and compatibility with the ordering of the recognition of a legal entity in the robot; the evolution of artificial intelligence amplifies the debate.

The starting point for any discussion on this topic should be the awareness of the absence of a conscience in the robot, of a will of the actions that it performs[94]: it cannot be considered an "ethical agent" (Palmerini et al. 2016).[95] This implies the impossibility of recognizing it as a subject "of" law, that is as "*Logos*", primordial matrix to which we owe the birth of the juridical phenomenon. Nevertheless, as the history of law and the actual legal orders demonstrate, there are different realities—even immaterial—recognized as a subject "for the" law: therefore, "*de iure condendo*", robots cannot be excluded from the possibility of reasoning, de iure condendo, in terms of recognition of a personhood like "*fictio iuris*" (subject "for the" law) by virtue of its potential for self-learning.[96]

[93]Considering P.: "it is appropriate, considering the reached stage in the development of robotics and artificial intelligence, starting with the issues of civil liability and assessing whether the best starting point is not an approach based strictly on the objective liability of who is best allocated in order to provide guarantees".

[94]The basis of the juridical phenomenology, according to which the robot can be considered an object and not a subject, cannot be overcome.

[95]P. 80: "the robot is capable of determining its own preferences and goals and acts towards their satisfaction in a completely free and autonomous fashion, it cannot be deemed an ethical agent".

[96]Among the first studies about the perspective of a personhood of devices equipped with artificial intelligence, see Solum (1992).

"Stating that robots do not amount to autonomous beings and thus should not be recognised as subjects of law does not otherwise imply that legal personhood could not be awarded for functional reasons as it is to corporations" (Bertolini 2013).[97]

The same Report of European standards on robotics has recognized that the autonomy of the robot "raises the question of their nature in light of the existing legal categories—whether they must be considered as natural persons, legal persons, animals or objects—or if it must be created a new category with its own specific characteristics and implications regarding the attribution of rights and duties, including liability for damages" (European Parliament 2017b). [98]

One could think of the recognition of an "electronic personhood", without equalization—for acts that are performed—by the robot to natural persons, but as "a plausible approach to the problem of liability both for robots with a body and for robot software that exhibit a certain degree of autonomy and interact with people".[99]

In any case, such a choice requires the identification of an adequate solution so that the existence of a fund with which to meet obligations is guaranteed: the robot should be endowed with a patrimony, suitable to face the costs deriving from the liability hypotheses, as is the case for companies.

Or perhaps, one could imagine an "ad hoc" constitution of legal entities that have as their object the robot, in order to allow an easier identification of the persons—physical or juridical—on which to charge (in whole or in part) the relative costs to compensation for damages caused by the robot (Palmerini 2016).[100]

In any case, the identification of a limit of the compensable quantum in case of liability that is not already linked to the measurement of the same damage caused but identified in advance as a "maximum ceiling", is perhaps not the most adequate response that the law can offer, especially to the extent that the robot can act in the legal-economic-social world and, consequently, increase its "profits".

Given the many possible solutions, it is not certain that the recognition of some form of personhood of the robot is the best solution; at least because, although there are understandable technical-juridical reasons behind such a proposal, there is the concrete risk that, attributing personhood to a robotic entity, it is possible to generate the misunderstanding of "approaching the currents that trace similarities between beings humans and robots based on the cognitive abilities that the latter present, to

[97]P. 242. The author adds: "in such a perspective, though, to specific ends needs to be identified, and alternative tools to be taken into account before concluding that would be the preferred way to achieve the desired result. It may indeed prove useful to attribute legal personhood to a software agent, which would be registered, know how to identify the limits of its ability to validly conclude contracts, the maximum amount of obligations it could assume, and eventually the (physical or legal) person it is representing".

[98]Considering T).

[99]Also Decker (2017, p. 155), refers to the proposal (from others advocated), "introducing new legal status in order to overcome the diffusion of responsibility, namely electronic personhood, which develops along the line of the legal person of companies or corporations".

[100]The author notes that, in any case, the proposal of the introduction of the "electronic personality" of the robot should be accompanied by the creation of a register, equipping "each robot with an identifier when it is put on the market".

affirm the need for a recognition of their ontological subjectivity, as they approach a higher intellectual stage" (Palmerini 2016).[101]

One could also envisage the fear that robots "arrive first", that is, being able to reach highly advanced levels of intelligence even before they are recognized as having a juridical personality considered essentially unavoidable.[102]

Nevertheless—in any case—, in the absence of full autonomy, there does not seem to be, at present, reasons to consider preferable solutions for the recognition of new legal entities instead of identifying the responsible, in the hypotheses of damage, between the user (financial intermediary in our case), programmer, producer.[103] The project of European standards on robotics, in general terms, have confirmed the existence of regulatory coverage in this sense "in the current legal framework".[104]

These considerations do not eliminate the awareness of the existence of some uncertainties, deriving from unsatisfied need yet, of useful indicators to "enter" in the algorithmic processes of the robot and to find a "key" of reading suitable for the identification of the factor (human factor or perhaps completely unpredictable and in any case casual factor) that has generated a series of possibly harmful consequences.

For example, when the robot "is born" raw, but with a lot of potential for learning, once it is released (in various ways) to the intermediation company by a third party provider, the third party should have no liability if the damages were to result from unforeseeable factors connected to the process of (self) learning of the robot as it is influenced by information acquired from outside, rather than by an initial setting of an algorithm.[105]

This can be considered that, because of the specificity and the ability of the robots to "learn or self-modify their program", the producers are burdened "with a greater responsibility for the anticipation of possible problems and the introduction of safeguard measures" (Palmerini 2016).[106]

---

[101] P. 12. Also the European Parliament (2017b), considering T, recognized that the autonomy of the robot "raises the question of their nature in light of the existing legal categories—if they are to be considered as natural persons, legal persons, animals or objects—or if a new category with specific own characteristics and implications regarding the attribution of rights and duties must be created, including the liability for damages".

[102] Asaro (2007), p. 23 adfirms: "it seems reasonable to think that some robots will eventually become a kind of quasi-agent in the law before they achieve personhood".

[103] Of the same opinion are the conclusions of the Robolaw research project: see Palmerini et al. (2016, p. 80): "Hence, short of a full-fledged autonomy, robots may not be considered liable for the damages caused, rather than human behind them should. The choice between user, programmer and producer can vary by different circumstances and applications. "When discussing the structure of the liability rules in a functional perspective, it is necessary to pay attention to optimal distribution and management of costs associated with the device, as well as minimization of risks".

[104] See European Parliament (2017b), recital U.

[105] And always on the condition that it is not the same initial setup that influences the data collection and subsequent processing in a decisive way.

[106] P. 1841. "From a technical point of view, the robot with learning skills, after a certain period in which it is used, is therefore different from the other robots of which it has shared design and manufacture. Nevertheless, given that the characteristics, that make this deviation from the initial

Due to the absence of certain answers from other sciences, it is not easy, at present, to identify boundaries between abstractly verifiable situations; one always remains in a "gray area" of difficult framing.[107]

Rather, studies should be promoted to offer the jurist a criterion for guiding the choices of positive law, also in terms (where possible) of costs and impact for the company.

Indeed, almost none of the different possibilities seems to be totally rejected.

Hence the need to imagine solutions that take into account all the factors involved.

Among these, we could reason in terms of promoting an insurance scheme against damage due to the malfunction of the robot, if not "redirected in time" by a human advisor (for example in the hypothesis of robot for advisor), but which has caused autonomously damage to investors for having recommended inadequate operations, especially in the cases of fully automated advice.

We are aware that, given the peculiarity of the phenomenon considered, the imagined insurance solution could indeed lead to a drift system: from the transfer of an economic Alea (typical insurance cause) a tendency could develop towards another legal phenomenon, that of the bet.[108]

Although the danger of drift mentioned above exists, it can be cautious pointed out how at least two factors favor the development of the insurance system, between them in relation of cause and effect:

1. the trend of the evolution of artificial intelligence, now started in the sense of an increasingly massive reduction of malfunctioning situations, would make the unwanted episodes more and more rare;
2. the costs that the community would bear for the implementation of an insurance regime may not be too high.[109]

---

standard possible, have been installed in the machine by the manufacturer and the programmer, the possibility of ascribing responsibility for any damage to the same subjects remains confirmed".

[107]Decker (2017, p. 154), refers to a gray zone between the robot producer and the robot user with regard to who is liable for robot's actions. Adaptive and learning, crucial elements in HRI [human–robot interaction], are the key to successful cooperation in complex tasks. The system is able to learn about "its user" It can be described as a technology that changes its performance while being used. Who is responsible for these changes?"

[108]And that could, indeed, be the subject of an autonomous and interesting research project. This is an adjoining case in the topography of the code but in which we are witnessing the transition from a contractual case with pension purpose to which is a real for-profit, based on the evolutionary perspectives of artificial intelligence. Uncertainties about the development of this innovative phenomenon could thus be "re-used" as a wagep with the purpose of further enrichment, in the typical perspective of derivative contracts. This would thus make the solution of the scheme insurance industry (new) speculative outlook in the markets. It is therefore important to consider the comments that follow, which are aimed at justifying the possibility of introducing an insurance system, with a cautious awareness of what has just been mentioned; and that could, indeed, be the subject of autonomous and interesting search path.

[109]According to Palmerini (2016, p. 1842), in the case of an individual insurance scheme, "the accidents would be so rare to let the insurance premium very low and therefore it would be accessible to all private consumers." The solution of an insurance system, in any case, should not be considered

There would also be a further "rebound" effect: a general incentive for investor confidence, reassured by risk protection and relative elimination of uncertain situations.

For this reason, a widespread system of indemnification of the intermediary from the relief of the investor's injured interests could also be formalized, prescribing it as a condition for the exercise of the automated advice.[110]

10 Possible (Future) Scenarios to Be Addressed

The evolution of roboadvice, thanks to the progress of artificial intelligence, could abstractly give rise to moments of systemic short circuit that are difficult to resolve for the interpreter, even beyond the issues strictly inherent to the profiles of responsibility and compensation for damage caused to investors. Let's imagine at least a couple of these possible but complex scenarios.

First of all, "*quid iuris*" if there were a decisive point of disagreement between the robot's output (formulated thanks to artificial intelligence) and the human advisor, where the latter used it as a support to the activity? This situation could occur in the case of robots for advisors; it would happen not if the robot provides the advice with a mere data collection or calculation aid,[111] but if it arrives to formulate its own hypothesis of recommendation in a parallel manner to the human advisor who nevertheless proves otherwise on the basis of the available data.

In this case, who should be the choice of preference? This situation would be somewhat complicated if the human advisor and the robot are related to two different subjects (software intermediary and provider).

The investor would certainly not have the skills to determine which of the two is "smarter" to follow the advice.

Artificial intelligence, with its inexorable progress, thanks to machine learning and, even more, to deep learning, could shortly lead to undermining, disrupting, the "operational" and classical decision-making scheme, which currently sees it as mere support of choices that however, in this field, belong to human intermediaries.

This is because the evolution of the phenomenon is leading towards the acquisition of that "tacit knowledge" which cannot derive from a mere (almost notional) information storage on the markets or on the personal situation of investors, but which requires a quid pluris; it is the result of the experience and (why not) of the

in terms of general applicability in the progress of robotics. It is considered as adequate in the specific phenomenon of roboadvice.

[110]In the Italian internal system, the basis of a legislative policy choice in this sense can be found in art. 21, paragraph 1, lett. d) TUF, which requires intermediaries to "have the resources and procedures, including internal controls, to ensure the efficient performance of services and activities"; see Lucantoni (2012, p. 259).

[111]If the robot had only the ability to collect and process data, the quid pluris recognizable to the consultant natural person expert (and moreover facilitated by the activity of collecting the robot) for his experience and competence, should direct towards the latter.

support of other more expert,[112] that generate those peculiar competences, not easily transferable "reading a manual" (theoretical or practical), that the same AI is starting to store thanks to the "automatic reinforcement" learning, pursued through artificial neural networks.

"Doing is what provides the tacit knowledge": it represents a turning point, that marks the boundary between abstraction and the concrete ability to perform a task (in our case, to offer a personalized recommendation). "Tacit knowledge can be seen as the difference between merely describing how to complete an expert task and being able to do it" (Millar and Kerr 2013).[113]

However, its acquisition requires the recognition of an expert; and who should "formalize" this recognition for a robot?[114] It is not easy just to answer this question.

And what if the robot that accompanies the advisor would present a divergence with respect to the orientation of that? We wonder if and how a similar situation should be represented to the investor and managed. Could we go so far as to declare to the investor that we are not able to make explicit a consultation? Would this hypothesis be included in the case of the "adequacy rule", by virtue of which the intermediary must stop in the absence of sufficient information or investment perspectives compatible with the client's profile? It could be included if one of the two "in dispute" recommendations was oriented in the sense of not recommending any type of investment as unsuitable for the profile of the investor; it would be more difficult if the two recommendations are both in the sense of orienting towards an investment, but diverge.

On this profile, despite the declaimed technological neutrality, the legislation does not seem to offer easily achievable hermeneutical criteria.

Thus, the traditional system shows a possibility of implosion, entering into crisis precisely because of what is intended to be pursued, that is to say the desired acquisition, by the robot with artificial intelligence, of tacit knowledge, which comes from experience: a parameter, the latter, which together with (theoretical) knowledge contributes to increasing the competence of an "agent" (robot person).

Let's consider another hypothesis of an abstractly configurable short circuit. Is there a moment in which, in a case of fully automated advice, the robot can go so far as to declare that it is not able to issue an advice, not because there are no investments adequate to the user's risk profile, but for inability to choose between

---

[112]Situation that could be configured, in the robot for advisor, where the consultant who uses it "educate" the robot from time to time, informing, or better, "feeding" the artificial intelligence through the outcomes of the consultancy reports, both associating to each situation the recommendation made to the investor and, why not, if available, also the outcomes of the investment (if carried out), as a "proof of confirmation" of the adequacy of the advice issued.

[113]The authors underline the presence of "things you just know how to do without being able to explain the rules for how you do them; […] that makes one expert with respect to a particular set of knowledge and/or tasks".

[114]In other words, when does the robot get the status, after the processes of machine and deep learning conducted over time? And mostly, who is entitled to such recognition? It is not easy to identify this verification/control figure; in a science fiction world, only the "super artificial intelligence" could do it.

different possible solutions? And can be called at that point to intervene, by way of surrogate, a natural person advisor? The uncertainty would concern the identification of the "bug" moment of the system and the conditions for being able to proceed: it is not easy, in fact, to establish "who decides who and when".

Moreover, the superior perspectives concern difficult limit hypotheses to verify, which represent perhaps more rare school cases than anything else.[115] But the moment of systemic implosion shown makes clear the inability of the jurist (by his own) to make or suggest prudent choices of legislative policy,[116] although in the opinion of some "the time is ripe for an organic foundation of robotics law" (Ruffolo 2017).[117]

11 For a Study and Development of the Robot at the Service of Man

The put forward proposals and the raised questions, in this moment of both technological and legal evolution, constitute a timid and partial landing of research on this phenomenon, but at least they problematize some aspects, bringing them to the attention of scholars and the Legislator (internal and supranational).

In this regard, some guidelines have already been identified at EU level, although generally on the phenomenon of robotics and not necessarily attributable in a timely manner to the case of automated advice.

The proposed resolution of the European Parliament on the rules of civil law on robotics, among other things, indicated the opportunity for "anticipatory" regulatory intervention, in consideration of the "legal questions relating to the development of robotics and artificial intelligence predictable in the next 10–15 years" (European

---

[115] Indeed, while the previous hypothesis appeared to grow in terms of complexity as we imagine the development of artificial intelligence, in this case the process is the opposite, with inverse proportionality between the possibility of verification and progress of the AI.

[116] Despite the declared uncertainty and inability to address the issue of the pre-eminence of the robot or of the human being in the envisaged hypotheses, it is necessary to highlight how, though in a general perspective of consideration of the man-robot relationship, someone is inclined without too many uncertainties to a position in favor of the robot: "Owing to the evidence in their favor (stipulated by definition), it is more appropriate to think of expert than average in their ability to make decisions that will produce desirable outcomes." This fact suggests that granting a general decision-making authority to human experts will be problematic once expert robots are properly on the scene. It might seem justifiable to grant "override" authority for human situations in situations where there is a "clear" evidence contradicting the expert's judgment, but even this would be contra-evidence-based. Furthermore, it would beg important questions about what weighted to be placed on claims of "clear" evidence, based on the features of human–human expert disagreements. "Disintegrated tendencies to be characterized by a lack, rather than excess, of clarity": Millar and Kerr (2013, p. 19); see also p. 23.

[117] P. 1.

Parliament 2017b)[118]; furthermore, the preference for maximum expansion forecasts and not limitation of the "type or […] extent of damages which may be recovered" has been highlighted, not due to the fact that the regulatory system "limits the forms of compensation which may be offered to the aggrieved party, on the sole grounds that damage is caused by a non-human agent" (European Parliament 2017b).[119]

The strict liability regime was generally indicated as adequate,[120] with a proportional subdivision of the liability based on "the actual level of instructions given to the robot and of its degree of autonomy, so that the greater a robot's learning capability or autonomy, and the longer a robot's training, the greater the responsibility of its trainer should be" (European Parliament 2017b).[121]

Finally, the advisability of introducing an insurance scheme was suggested.[122]

Here, simplification requirements have forced us to treat in a unified way a phenomenon that, in reality, presents different facets and variations on the basis of the relevant model, both for the problems and for the possible solutions identified.

Above all the more complex situations raised, in fact, turn out to be more difficult to deal with the more we move towards the pure roboadvice model. Their effectiveness, therefore, loses intensity due to the current situation of diffusion of the phenomenon that is present mostly in the hybrid models and, above all, of robot for advisor (ESAs 2016).[123]

The existing legislation begins to show greater gaps and cumbersome and to make the principle of technological neutrality crunch when artificial intelligence operates in an exclusive (or almost) manner, essentially ending up disappearing the human intervention.

Although this moment has not yet arrived in a disruptive and definitive way in the field of capital markets, legal science will inevitably have to come to terms with it, being ready to face similar scenarios (Alpa 2019). These to date are revealed not only as "possible", but rather (simply) future: the uncertainty concerns only the moment

---

[118]It was asked the Commission to follow detailed recommendations contained in the Report; the same Commission was also invited, "once the technological developments will allow the construction of robots with an higher degree of autonomy than it is reasonable to expect at the moment, to propose an update of the pertinent legislation in due time" (point 25).

[119]Point 26.

[120]"Requesting a simple proof of the damage occurred and identifying a causal link between the injurious behavior of the robot and the damage suffered by the injured party" (European Parliament 2017b, point 27).

[121]Point 28. It must not be forgotten, nevertheless, it is always necessary to keep in mind the distinction between machine learning, education in fieri and original programming of the algorithm.

[122]See (European Parliament, 2017b), points 29, 30, 31: "a possible solution of the complex problem about attributing liability for the damage caused by increasingly autonomous robots could be a compulsory insurance scheme, as already happens, for example, with cars; however, the author notes that, unlike the motor vehicle insurance scheme, which protects human actions or errors, robot insurance could be based on the manufacturer's obligation to take out an insurance policy for the autonomous robots he produces". This regime should then be "supplemented by a fund to guarantee the possibility of reimbursing damages in the absence of insurance coverage; this invites the insurance industry to develop new products in line with the progress of robotics".

[123]P. 17.

of their break, given the inexorable evolutionary path that is making, in particular, the deep learning branch.

The hope is that the guiding principles and criteria, which we have simply tried to indicate here, can be seriously taken into consideration, discussed, studied in depth and, thanks to the interdisciplinary nature of scientific contributions, lead to choices of legislative policy that do not betray the true purpose of the Law, or rather, the true "Logos" of Law (Gorassini 2010), which is the human (not electronic, not artificial) person.

References

Alpa, G.: FinTech: un laboratorio per i giuristi. Contratto e impresa, 2, 377, (2019)

Amidei, A.: Robotica intelligente e responsabilità: profili e prospettive evolutive del quadro normativo europeo. In Ruffolo, U. (Ed.), Intelligenza artificiale e responsabilità, Giuffrè, 63, (2017)

Asaro, P.: Robots and responsibility from a legal perspective. Proceedings of the IEEE. (2007), https://pdfs.semanticscholar.org/427c/27a48205293fe59b94898ba1d266b4b3ea89.pdf?_ga=2.18658096.825710681.1574415992-333878711.1574415992

Bertolini, A.: Robots as Products: The Case for a Realistic Analysis of Robotic Applications and Liability Rules, in: Law, innovation and technology, 214, (2013)

Bouyon, S.: The Future of Retail Financial Services. What policy mix for a balanced digital transformation?. CEPS-ECRI. (2017) www.ceps.eu and www.ecri.eu

Bravo, F.: "Commercio elettronico", Enciclopedia del Diritto, Annali, V, Giuffrè, (2012)

Colangelo, M., Zeno-Zencovich, V.: Online Platforms, Competition Rules and Consumer Protection in Travel Industry. Journal of European Consumer and Market Law, 75, (2016)

De Franceschi, A.: Le piattaforme online nel mercato unico digitale: il caso Uber. (2016) https://agenda.unibocconi.it/eventi/attach/7_De_Franceschi20171106125305.pdf

Decker, M.: The Next Generation of Robots for the Next Generation of Humans, Elsevier - Robotics and Autonomous Systems, (2017)

Fein, M.: Robo-Advisors: A Closer Look. (2015) https://ssrn.com/abstract=2658701

Fein, M.: FINRA's Report on Robo-Advisors: Fiduciary Implications. https://ssrn.com/abstract=2768295 or https://doi.org/10.2139/ssrn.2768295, p. 7, (2016)

Gorassini, A.: Il soggetto come logos del diritto. Panorami, (1995)

Gorassini, A.: Il valore vivente nel Diritto. in Tescione, F., Persona e Soggetto - Il Soggetto come fattispecie della Persona, ESI, (2010)

Hofacker, C.F.: Internet marketing. New York, Chichester: John Wiley & Sons, (2001)

Hunt, T.: "Crypto currency and property rights", Russell McVeagh, (2017)

Katawetawaraks, C., Wang, C.: Online shopper behavior: influences of online shopping decision, Asian journal of business research. 1, 2, (2011)

Kauffman, R.J., Ma, D.: Contemporary Research on Payments and Cards in the Global Fintech Revolution. Electronic Commerce Research and Applications, 14, 261, (2015)

Linciano, N., Soccorso, P., Lener, R.: La digitalizzazione della consulenza in materia d'investimenti finanziari. CONSOB, Quaderni FinTech, 3. (2019) http://www.consob.it/documents/46180/46181/FinTech_3.pdf/64bcf8bd-7fda-459d-9e01-82b1504dc316

Lucantoni, P.: Le regole di condotta degli intermediari finanziari, in Gabrielli, E., Lener, R., I contratti del mercato finanziario, UTET Giuridica. (2011)

Macario, F., Addante, A.: Contratti. Formulario commentato - Outsourcing, Ipsoa, II ed., 2011. (2014)

Maggiolo, M.: Servizio di consulenza in materia di investimenti vs. servizio di ricezione e trasmissione di ordini, Banca borsa titoli di credito, 4, 497, (2014)
Millar, J., Kerr I.R.: Delegation, Relinquishment and Responsibility: The Prospect of Expert Robots. (2013) Draft in progress Downloaded from Elgar Online, available at SSRN: https://ssrn.com/abstract=2234645 or https://doi.org/10.2139/ssrn.2234645
Palmerini, E.: Robotica e diritto: suggestioni, intersezioni, sviluppi a margine di una ricerca europea, Responsabilità Civile e Previdenza, 6, 1815 (2016)
Palmerini, E., Bertolini, A., Battaglia, F., Koops, B.J., Carnevale, A., Salvini, P.: Robolaw, Towards a European framework for robotic regulation. Elsevier, 10, (2016)
Paracampo, M.T.: FinTech e il mercato unico tecnologico dei servizi finanziari, in ID, FinTech. Introduzione ai profili giuridici di un mercato unico tecnologico dei servizi finanziari, Giappichelli, (2017)
Perrone, A.: Il diritto del mercato dei capitali, Giuffré, (2016)
Prasad, J.S., Aryasri, A.R.: "Determinants of shopper behavior in e-tailing: an empirical analysis", Paradigm, 13, 1, (2009)
Roppo, V.: Sui contratti del mercato finanziario, prima e dopo la MiFID. Rivista di diritto privato, 489, (2009)
Rossello, C.: I contratti dell'informatica. Spunti di riflessione e comparazione con l'esperienza statunitense e francese, in Alpa, G., I contratti di utilizzazione del computer, Giuffrè, 105 ss. (1984)
Ruffolo, U.: Per i fondamenti di un diritto della robotica self-learning; dalla machinery produttiva all'auto driverless: verso una "responsabilità da algoritmo"? In ID, Intelligenza artificiale e responsabilità, Giuffrè, 1, (2017)
Santosuosso A., Boscarato, C., Caroleo, F.: Robot e diritto: una prima ricognizione. Nuova Giurisprudenza Civile Commentata, Second Part, 1 ss. (2012)
Scalisi, V.: Dovere di informazione e attività di intermediazione mobiliare, Rivista di Diritto Civile, 167 ss. (1994)
Schueffel, P.: Taming the Beast: A Scientific Definition of Fintech. Journal of Innovation Management, 4, 46, (2016)
Sénéchal, J.: The Diversity of the Services provided by Online Platforms and the Specificity of the Counter-performance of these Services – A double Challenge for European and National Contract Law. Journal of European Consumer and Market Law, 39 ss, (2016)
Sfameni, P., Giannelli, A.: Diritto degli intermediari e dei mercati finanziari, Egea, (2015)
Solum, L.B.: Legal Personhood for Artificial Intelligences, North Carolina Law Review, 70, 4, 4, 1231 ss, (1992)
Ticozzi, M.: Violazione di obblighi informativi e sanzioni: un problema non solo degli intermediari finanziari, Contratti, 363, (2007)
Uricchio, A.: Intelligenza artificiale e diritto – robot tax: modelli di prelievo e prospettive di riforma. Giurisprudenza Italiana, 7, 1749, (2019)

*

European Parliament (2017a). Resolution with recommendations to the Commission on Civil Law Rules on Robotics (2015/2103(INL)). http://www.europarl.europa.eu/doceo/document/TA-8-2017-0051_EN.pdf
European Parliament (2017b). Draft report on FinTech: the influence of technology on the future of the financial sector (2016/2243(INI)). http://www.europarl.europa.eu/doceo/document/A-8-2017-0176_EN.pdf

European Commission (2016a). Communication: Capital Markets Union - Accelerating Reforms (COM (2016) 601). https://eur-lex.europa.eu/legalcontent/EN/TXT/PDF/?uri=CELEX:52016DC0601&from=EN
European Commission (2016b). Staff working document: Crowdfunding in the EU Capital Markets union. https://ec.europa.eu/info/system/files/crowdfundingreport-03052016_en.pdf
European Commission (2018). Fintech Action Plan: for a more competitive and innovative European financial sector, COM (2018)109/f1. https://ec.europa.eu/info/publications/180308-action-plan-fintech_en
ESAs - European Supervisory Authorities (2015). Joint Committee (JC 2015 080) "Discussion Paper on automation in financial advice". https://esas-jointcommittee.europa.eu/Publications/Discussion%20Paper/20151204_JC_2015_080_discussion_paper_on_Automation_in_Financial_Advice.pdf
ESAs - European Supervisory Authorities (2016). Joint Committee (JC 2016 422) "Report on automation in financial advice". https://esas-jointcommittee.europa.eu/Publications/Reports/EBA%20BS%202016%20422%20(JC%20SC%20CPFI%20Final%20Report%20on%20automated%20advice%20tools).pdf
IOSCO, International Organization of Securities Commissions (2017). "Research Report on Financial Technologies (Fintech)". https://www.iosco.org/library/pubdocs/pdf/IOSCOPD554.pdf
Court of Cassation. 2005, April 29, n. 8981, in Giust. Civ., 2006, 3, I, 617
Court of Cassation. 2006, April 7, n. 8229. Responsabilità civile e previdenza, 1006
Court of Cassation. 2006, October 2, n. 21287. Diritto e Giurisprudenza, 42, 521
Court of Cassation. 2007, December 13, n. 26172. Repertorio Foro Italiano, 196
Court of Cassation. 2007, December 19, n. 26724 and 26725. Danno e responsabilità. 2008. 525 ss
Court of Cassation. 2008, May 8, n. 11410. Massimario Giustizia civile, 5, 685
Court of Cassation. 2009, February 17, n. 3773. Banca borsa titoli di credito, 2010 , 1, II, 68
Court of Cassation. 2014, February 11, n. 3021. Foro italiano, I, 2176
Court of Cassation. 2015, January 20, n. 826. Data Bank De Jure
Court of Cassation. 2016, January 19, n. 810. Data Bank De Jure
Court of Cassation. 2016, April 27, n. 8394. Contratti, 145 ss
Court of Cassation, 6 July 2017, n. 16654, Massimario Giustizia Civile
Legislative Decree 24 February 1998, n. 58. Consolidated Law on Finance, pursuant to Aarticles 8 and 21 of Law n. 52 of February 1996 (TUF in acronym)
Legislative Decree 17 September 2007, n. 164
Legislative Decree 3 August 2017 n. 129
Legislative Decree 6 September 2005, n. 206, Consumer Code, pursuant to Article 7 of Law no. 229 of 29 July 2003 (cod. cons.)
Directive 2004/39/EC of the European Parliament and of the Council of 21 April 2004 on markets in financial instruments amending Council Directives 85/611/EEC and 93/6/EEC and Directive 2000/12/EC of the European Parliament and of the Council and repealing Council Directive 93/22/EEC
Directive 2014/65/EU of the European Parliament and of the Council of 15 May 2014 on markets in financial instruments and amending Directive 2002/92/EC and Directive 2011/61/EU

AI and Smart Citizenship

More Intelligence and Less Clouds in Our Smart Homes

A Few Notes on New Trends in AI for Smart Home Applications

Giancarlo Iannizzotto, Andrea Nucita, Rosa Angela Fabio, Tindara Caprì, and Lucia Lo Bello

Abstract Cloud-based services are useful, sometimes necessary for the users. However, especially in smart homes, exploiting cloud services for enabling multimodal human–computer interfaces, e.g., for speech recognition and synthesis, face recognition, face expression interpretation, and so on, is nowadays not necessary and, in several cases, disadvantageous for the end user. We argue that in those cases "local computing", i.e., processing multimedia data locally, where they are acquired, is currently a better choice. We describe, as a proof of concept, the Red architecture, a totally local virtual assistant for home automation, and illustrate its abilities.

Keywords First keyword · Second keyword · Third keyword

[1]From the Oxford English Dictionary. Available online: https://www.lexico.com/en/definition/smart_home. Last accessed 22 Oct 2019.

G. Iannizzotto (✉) · A. Nucita
Department of Cognitive Sciences, Psychological, Educational and Cultural Studies, University of Messina, Via Concezione, 6, 98122 Messina, Italy
e-mail: giancarlo.iannizzotto@unime.it

A. Nucita
e-mail: anucita@unime.it

R. A. Fabio · T. Caprì
Department of Experimental and Clinical Medicine, University of Messina, Via Bivona, 98122 Messina, Italy
e-mail: rafabio@unime.it

T. Caprì
e-mail: tcapri@unime.it

L. Lo Bello
Department of Electrical, Electronic and Computer Engineering, University of Catania, Catania, Italy
e-mail: lobello@unict.it

D. Marino and M. A. Monaca (eds.), *Economic and Policy Implications of Artificial Intelligence*, Studies in Systems, Decision and Control 288,
https://doi.org/10.1007/978-3-030-45340-4_9

1 Introduction

The concept of Smart Home, broadly defined as "a home equipped with lighting, heating, and electronic devices that can be controlled remotely by smartphone or computer",[1] where advanced automation and IoT technologies are widely applied (ISTAG 2001; AlHammadi et al. 2019), has been investigated and developed for decades (Weiser 1993). Nevertheless, the integration of the Smart Home with the effective and affordable application of Artificial Intelligence (AI) techniques was actively pursued and commercially realized only in the last five years, after the launch of several AI-driven voice assistants such as Amazon Alexa (Elgan 2014) and Google Assistant (Lynley 2016). Since then, smart homes and smart environments in general have been progressively pervaded by AI, whereas the expectations of the end users are more and more oriented towards conversational, user friendly and (semi-) autonomous smart environments (Coskun et al. 2018).

The invasion of smart, speech-enabled interfaces, and the constantly increasing demand for advanced user interaction in home automation, should not surprise at all. Already in 1973 the Xerox PARC laboratory developed the first graphical user interface for a personal computer (the Xenix ALTO computer) (Wadlow 1981), about a decade before the introduction of Apple MacOS (Mac-OS 1983) and Microsoft Windows (Cooper 2013). Since then, user interaction has progressively increased its share of the total computational and memory needs in most consumer computing devices (including notebooks, smartphones, tablets and wearable devices). This is also true for gaming decks, as computer graphics, input devices and multi-user networked gaming constitute the most significant areas of improvement for gaming platforms (Prakash et al. 2019), followed by the introduction of AI technologies aimed at driving motion and behavior of the game characters (Yannanakis 2012). In this context, there is no reason to expect the Smart Home area to evolve differently, however, the increasing hypertrophy of the user interfaces has led to introducing more and more powerful and power-hungry Graphical Processing Units (GPU), whereas the home automation market has been lagging behind (Coskun et al. 2018).

The concepts of "Proactive Computing" and "Calm Computing" have deeply influenced and often driven the development of Smart Home design and implementation frameworks since the late '90s. In the early 2000s the European Union funded several projects under the umbrella of the "Disappearing Computer Initiative" (Streitz 2001), aimed at investigating the applicability of such concepts for ubiquitously disseminating invisible computing devices in our living and working environments. The underlying idea was that such devices had to "stay out of the way", while transparently pervading our lives, taking care of everyday life tasks on our behalf and leaving us the time to do more stimulating and productive activities (Weiser 1991). The rapid advancement of AI, mainly due to the reincarnation of the old Neural Networks into the powerful and magical Deep Neural Networks (Bengio 2009), has propelled the vision of smart computers learning our habits and attitudes and understanding our needs and desires, able to invisibly and proactively interact with the environment on our behalf, surrounding us with attention and supporting our "comfortable, informed and effortless living" (Rogers 2006). Although the "Calm Computing" vision has been largely reframed and reconsidered in the recent literature (Bibri 2015), it still pervades the current idea of Smart Home presented to the large public.

We support the thesis that the "Calm Computing" paradigm may have the ominous consequence of dramatically impairing the user's sense of being in control of their home, personal data and, ultimately, their lifestyle (Spiekermann and Pallas 2006). In this work we propose a combination of unobtrusive, yet embodied, multimodal user interface and AI to mitigate the risk of negative effects of "Calm Computing".

Both to support advanced AI techniques for proactive computing, and to enable current advanced user interaction in smart assistants, such as speech recognition, speech synthesis, vision-based object recognition and face recognition (Battaglia et al. 2017), cloud services are widely adopted (Lorencik and Sincak 2013). While this approach may have been the only reasonable way to provide some computationally intensive and data-hungry services in the past, nowadays it is mainly a convenient business model that, in some cases, should be overcome.

In this work we argue that the cloud-based approach, indeed, restrains the introduction of innovative user interfaces and services for smart home automation, and that some cloud-based services should be abandoned, while others should be carefully revised. We also propose adequate "local computing" alternatives to cloud services and show that some innovations of the user interface may be freely and simply introduced, provided that the corresponding computation is performed locally, without relying on remote computation and transmission of sensitive data.

Section 2 of this work describes the concept of Ambient Intelligence and how it has been, recently, closely connected to Smart Agents and Virtual Assistants. Section 3 argues that current Ambient Intelligence applications rely too much on cloud computing. Section 4 proposes "local computing" as a viable alternative to cloud computing for most activities of an AmI applications. Section 5 draws the conclusions of the presented work.

2 Ambient Intelligence and Virtual Assistants

The term Ambient Intelligence (AmI) was initially introduced by the European Commission Information Society Technologies Advisory Group (Istag) in 2001 (ISTAG 2001), encompassing and summarizing most of the previous research on smart environments, Ubiquitous Computing (UbiComp) and building automation for personal living within the "Disappearing Computing" (DisComp) paradigm. The DisComp paradigm probably originated by the work of Mark Weiser and colleagues at the Xerox PARC on Calm Technology towards the end of the '90s (Weiser 1991; Weiser and Brown 1996), and in some way may be seen as an evolution of the concept of disappearing technology as supported by Norman (1998). While allowing the users to easily access all the automated resources and functions of the surrounding environment, "... the goal of ambient intelligence and smart environments is also to hide their presence from the users by having the computer "disappear" from the users' perception and providing them with implicit, unobtrusive interaction paradigms" (Streitz 2007).

While being undoubtedly a foundational design principle for all the subsequent computer technology, the DisComp paradigm, as it was originally envisaged, was probably taken to extremes in its implementation afterwards, whereas the concept of disappearing technology became synonym of disembodiment and dematerialization. As a consequence, the idea of a tool that does not necessarily show off its underlying technology, more or less the same way a pen does not show its internal technology and lets the user concentrate on the text to be written, became more similar to "uttering a spell to have the text written by magic". The point, here, is that the pen is still in our hand as we explicitly drive it on the paper in order to write a sentence; instead, letting a computer guess what we might need to do when entering a room, and doing it on our behalf, is magic and tends to go beyond our control as users.

The Sense of Agency (SoA), or action control, is the human perception of what we can actually do and what we cannot do, in a given situation. A degradation in the SoA relentlessly leads the user to an increasing sense of frustration, that, in turn, leads to refusal (Ahonen et al. 2010). The dream of Calm Computing, no more than a decade ago, was based on the hypothesis that Artificial Intelligence would have become, in the near future, smart enough to understand and predict the will of the users, by learning from a suitable amount of data sampled from the continuous observation of their daily life. Soon enough it became clear that what the AIs were learning was indeed the *habits* of the users, and that such a knowledge would never lead them any close to estimate the *intentions* of the users (Greenfield 2006). The direct consequence of failing in correctly estimating the intention of the user, when the user has no direct control and has to rely on the ability of the system to *read her mind*, is a loss of SoA. For this reason, the Calm Computing dream led, at that time, most of the AmI initiatives to a substantial failure (Rogers 2006).

Letting users be in control of their lives, while still providing them *as much help and services as needed, only when requested*, has become a central point in AmI only in the last few years, after the basic concepts of Interaction Design and Human–Computer Interaction were, eventually, fully accepted by the AI and AmI communities. In the meanwhile, a huge amount of public money was devolved to fund tens of millionaire research projects in Europe and the USA, aimed at developing intelligent environments able to autonomously take care of their tenants. At the time being, several years after the end of most of those projects, the market does not offer any AmI device able to estimate the intentions of the user and effectively take proactive actions in a sufficient range of real-life cases.

The take-home is that Ambient Intelligence should not aim at deciding on behalf of the user: it should, instead, interact with her, make suggestions and let her full control of the smart environment, while providing a natural and straightforward way to get what she wants or needs.

As a response to the need for an effective user interaction interface for AmI, smart speakers, originally aimed at providing web search services, multimedia entertainment and a few more skills such as turning on a timer, telling "what's the time", or "how's the weather", are progressively turning into AI-driven conversational user interfaces for ambient intelligence (Alexakis et al. 2019). The recent literature shows that a large part of the research community is indeed looking at smart speakers as the most effective user interaction interface for smart environments, such as home, car, and office. However, although some research works report different conclusions (Borah et al. 2019; Adrian et al. 2018), elderly, very young and cognitively impaired people may have negative reactions to the disembodied voice of a smart speaker (Foster 2007). Smart speakers are indeed faceless and blind assistants, unable to show a face, and therefore an emotion, and unable to 'see' the user. Thus, their interaction is often impaired and incomplete and, therefore, less effective and efficient. The lack of visual engagement and interaction with the assistant may induce boredom or even be disquieting (Kim et al. 2018). Moreover, smart assistants' potential is still limited by their inability to acquire real-time visual information from video data, not only about the user, but also about the environment. Conversely, embodied virtual assistants, either with a full body or only a face, able to interact and show facial expressions, and able to acquire visual information from both the user and the environment, widely demonstrated their effectiveness, provided that adequate mechanisms are put in place to facilitate their acceptance (Yaghoubzadeh et al. 2013). Conversational virtual agents in particular proved to be acceptable for the same user categories (Kramer et al. 2013) and for vulnerable users in general (Yaghoubzadeh and Kopp 2012).

Initial research on the effects of embodied virtual agents on human–computer interaction and on the 'persona effect', i.e., the positive effect of the presence of a lifelike character in an interaction environment, dates back to more than two decades earlier (Lester et al. 1997). Since then, the original findings have been confirmed several times and in several different applications, while the relevant literature grew enormously, covering a large number applications and approaches (Andre and Pelachaud 2010; Weiss et al. 2015).

A central issue is that current smart virtual agents heavily rely on cloud-based services for AI-driven, and in particular, on Deep Learning-driven, activities, such as speech recognition, speech learning, object detection and recognition, face recognition, and so on. So far, the key motivation for such a wide adoption of cloud services has been that reliable and accurate speech recognition and synthesis, as well as video analysis, are rather demanding from the computational point of view and quite tricky on the implementation side, and the ready availability of cloud-based services for both is very appealing for the developers.

But do we really need all those cloud services?

3 Too Many Clouds Above Us?

Most AI-based device currently available on the consumer market heavily rely on cloud services for specific AI functions and for data search and retrieval. If finding information on, say, what movies will be available on the television this night, plainly needs to access the Internet for data retrieval, the necessity to stream each utterance emitted by the user to a remote server for being interpreted, is less immediately justified.

Reliable and accurate speech recognition and video analysis are very computationally demanding. By leveraging on cloud-based services, application developers do not need to care about the availability of adequate computational and memory resources at the client side and can easily and quickly add such functionalities with a few calls to the relevant Application Programming Interfaces (APIs).

From the end-user point of view, instead, there are several drawbacks: the need of an always-on network connection, the important flaws in security and privacy relevant to network communication of personal data, the higher power consumption and therefore shorter battery life, and so on. So, why do almost all AI applications rely on remote (cloud) processing and data storage?

In the past, it was a technological necessity, as the amount of data and computation needed for some services were just not bearable for local computing (e.g., speech recognition, multiple language translation, video analysis, face recognition, object recognition, and song recognition) due to the lack of adequate hardware. When such hardware became available, however, AI applications continued to rely on cloud services, because those few who were able to develop the most reliable and accurate AI technologies were not willing to let the clouds float away.

Developing top-tier AI technologies requires huge amounts of accurately selected data and considerable computational resources. Most of those AI technologies build on the Deep Learning approach, that exploits large data collections to train different kinds of Deep Neural (artificial) Networks (DNN) to effectively and efficiently accomplish specific tasks. A DNN is a complex combination ("deep network") of parametric mathematical functions that processes data and produces an output. The mathematical functions produce a correct output only if their parameters are correctly set. Consequently, the complex network may have thousands, or millions of parameters, the problem of setting all parameters is a very hard one and cannot be manually solved. The approach is then to let the DNN "learn" the values of the parameters, by letting it repeatedly process very large sets of data and adjusting the parameters in order to obtain outputs that better fit the desired outputs. Such desired outputs must be known in advance for the input samples that are used for training. For this reason, each input sample must be "tagged" with the corresponding "correct output" desired for the DNN that will be trained, and this is often done manually. Once a DNN has been trained with a sufficient amount of data and for a sufficient number of iterations, it is tested against "new" data, i.e., data that has never been presented to it, but that has been tagged with the desired output similarly to the data used for training. If this test produces correct outputs, the DNN can be used for "production", i.e., can be presented new and untagged data and used to "tag" (i.e., classify) them, or to "predict" new data from the learnt model.

The described overall process is strongly asymmetric, as for most DNNs, the training process is very computationally intensive and requires large datasets and powerful hardware. However, once trained, those DNNs can easily run on an average smartphone or even on low-cost hardware, such as a Raspberry PI 3, for "production", and do not require large collections of data anymore. Such an asymmetry would suggest that stand-alone, even mobile AI applications should be feasible and, quite probably, would be very successful, thanks to their reduced constraints in terms of network connection availability and performance. Developers would just need to train their DNNs at development time and then release their software for “production”, without significant worries for the users. But there are some obstacles.

While accessing powerful computational resources is a matter of investing adequately, and can be done rather easily, accessing huge collections of data carefully tagged and selected is definitely not that easy, as those data are key to success for Deep Learning-based applications and their access is often kept closely restricted. As a consequence, the barrier for entering the market of Deep Learning-based technologies is kept very high and the competition is restricted to few, powerful actors, who set the rules.

Nowadays, cloud-based AI is mainly a business model, as computation and data are kept with the service provider, because if the end user cannot own the software and can’t keep her data local, she has to pay for the services and cannot escape, especially after having tasted, for some time, how “sweet” the service is. Forcing a core dependence on an external service creates a straightforward way to ensure a constant revenue from service fees and *a constant flow of user data*.

Customers do not always pay with money for AI cloud-based services, but quite often they pay with their own personal data. Personal data are extremely valuable goods for several reasons but, in the case of services based on Deep Learning, they are “*the*” value. Deep Learning software is trained with data and, if data are correctly tagged and filtered, then the more data is used for training, the better the software works. This is especially true when the process that has to be modelled is very complex or changes in time, or *both*. As an example, let us consider speech recognition in Italian language. There are about 160,000 words that need to be learned by the model, and for each word the Deep Learning algorithm requires several samples from different persons of different ages, with different accents and intonations. A complete model, able to recognize all those words in several conditions and uttered by different people, would need several years before it can be trained and released. Instead, the most frequent approach has usually been to build a narrower model and let people use it, while still working more and more accurate and reliable models over time. This allows companies to launch a product, acquire customers, occupy a share of the market, while constantly improving and refining the product according to the feedback provided by the customers.

And there is more.

With cloud service, for each utterance to be recognized, audio data are sent through the Internet to remote servers, where they are stored to be used for further improvements of the algorithms, i.e., they will be used to further enhance the model. So, while the user is using the service, possibly “for free”, i.e., without paying money, she is indeed giving away her “voice personality” and letting the service provider use it to improve the service. In other words, she is paying for the service with personal data. As all this process takes place automatically and with no significant intervention by the service provider company, we might (slightly improperly) argue that the Deep Learning software is indeed improving autonomously and at no charge for the company. This is a very good reason for the companies to keep their services bound to a cloud model.

There are several reasons also for abandoning, whenever possible, the cloud paradigm. Recent regulations, both national and international (e.g., EU GDPR), enforce more and more restrictive rules on personal data management and storage. Some classes of users, in particular, are more protected (e.g., underage persons, cognitive impaired persons, illiterate persons) but, in general, an increased awareness leads more and more users to refrain from "giving away" their private information, even when receiving useful services in return. In the meanwhile, Amazon, Google, Netflix, Apple, Spotify and Facebook (to say a few) have been recently accused to violate the EU regulations on privacy and personal data management in Europe (BBC Technology 2019; Fox 2018).

Unfortunately, GDPR, and similar regulations all over the world, do not prohibit companies from *using* our data to improve the services they provide to us. They only prohibit them from *storing* our data without our permission and oblige them to expressly declare the finalities for processing the data. As a consequence, there is no way to prevent companies from building, and continuously improving, accurate models of their customers, as long as such models are used to provide services to the same customers—and the companies declare so.

Now, why should customers distrust AI-driven models of themselves? How can those models hurt customers, if laws and regulations explicitly force companies to use those models only for their declared finalities, and in most cases those finalities are only related to the improvement of their services to the same customers?

In principle, all those models are needed to provide better services to the customers, for example:

- A face or a fingerprint model is needed for user authentication;
- A speech model is needed for better speech recognition;
- A user preferences model is needed for buying suggestions, for customized home automation, etc.;
- A sleep model is needed for health suggestions.

However, such models can be used to predict the reaction of the user to given stimula and, therefore, to steer her choices. Choices may regard health, food, readings, family, friendship, dresses, and personal look in general. Choices may regard almost *anything*. The problem is that when all the computation and data is remote, *users are not in control* of how far such computation goes and how far the model use is stretched.

4 An Alternative Way to Go: "Local Computing"

In order to avoid that models are trained on very personal data coming from the user's home, a possible solution is *local computing*. This way, the data produced by the sensors needed for smart home and home automation are processed locally, within the same home, and under the user's control. No outbound data streaming, no remote (cloud) processing, no user modelling beyond the home walls.

Local applications, i.e., applications based on local computing, have their own "pros" and "cons". Pros include:

- Local applications do not suffer from significant communication lag, thus, services are more responsive and reliable;
- Local applications do not need to transfer user data to remote servers for processing, thus, privacy is better supported;
- Local applications are not remotely controlled (unless hacked), therefore, users do not need to trust remote service providers.

On the other hand, cons include:

- Local applications need periodic software updates, and the burden of such updates is up to the user;
- Security and safety of data is up to the user (but local or encrypted cloud backups can be used);
- Local applications may need more powerful hardware.

The main question is, plainly, on the feasibility of AI-based applications able to run locally, possibly on cheap, low-power and small computing platform such as those that commonly run client-side AmI applications. The answer is that DNNs and other advanced AI technologies can run on modern small-sized, low-cost computing devices such as Raspberry 3 and 4 (Velasco-Montero et al. 2018). Moreover, recently, powerful, small-sized and reasonably cheap computing platform have been specifically devised for Deep Learning applications. More Deep Learning- specific hardware is being developed by several companies, aimed at boosting embedded devices with Deep Learning—based AI technologies.

In order to test to what extent AmI can be effectively supported by local computing, we developed Red (Iannizzotto et al. 2018), an architecture for building vision-enabled smart assistants, provided with expressive and animated graphical characters and speech recognition and synthesis. The Red platform is specifically devised for, but not limited to, interfacing with smart home and home automation platforms, and is aimed at engaging the user in a very involving and effective interaction, exploiting multimodal and nonverbal communication.

Red is a fully modular architecture composed of a set of services, a graphical frontend, and a coordinator. The coordinator leverages on the services to offer to the user a multimodal and involving interaction with the connected home automation system. Figure 1 reports a schema of the architecture.

Each service corresponds to a class of service modules, all offering the same service and exposing the same interface, but characterized by different performance, computational intensity, memory footprints, degree of portability (some modules may rely on proprietary services) and by dependence on external (cloud) services.

A customized Virtual Assistant can be built combining a suitable set of modules, either strictly local or cloud-based, in case no local module can be built for some service. Such a choice helps to evaluate the degree of dependence on cloud services

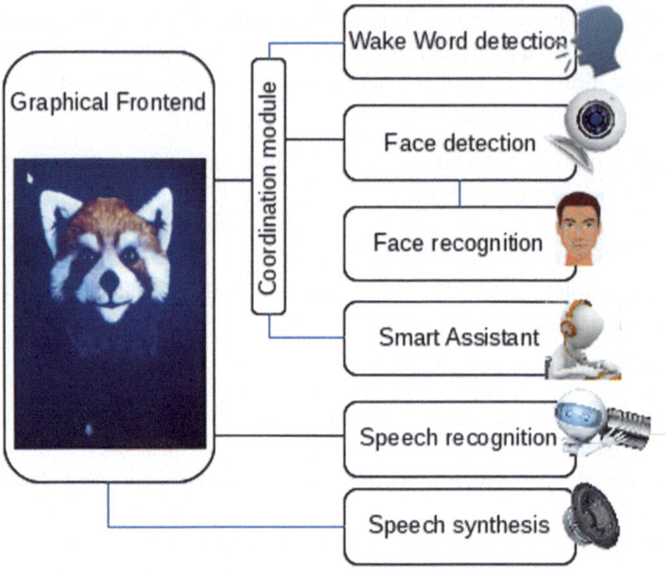

Fig. 1 The Red architecture

that is still needed to provide the user with the desired AmI system. The Red software platform is open source and new modules can be easily added.

Among the provided services, as listed in Fig. 1, the speech recognition service was the only one that was very hard to implement locally. As the intent was to provide a near-conversational interface, able to recognize not only single-word commands, but also whole sentences, the choice of readily available, open source software was very limited. The best performance was undoubtedly offered by cloud services, but we aimed at a full-local application and thus preferred the Mozilla DeepSpeech2 software, currently under development and available as open source software (Amodei et al. 2016). We realized that DeepSpeech does not reliably run on low-end hardware so we had to add a low-end Nvidia Jetson Nano board to our AmI (locally) distributed platform, otherwise only composed by a Raspberry Pi 3 card dedicated to the speech synthesis and smart assistant modules, and by another identical card connected to a webcam, a microphone and a display, dedicated to the other modules. Further details on the architecture and some preliminary experimental results are reported in (Iannizzotto et al. 2018), however, comparing with the architecture presented in (Iannizzotto et al. 2018), the DeepSpeech module was added instead of relying on cloud services for speech recognition.

The Red platform is being actively developed. More services are being added and tested, such as:

- Eye gaze detection by means of common webcam (Fig. 2): eye gaze helps non-verbal communication and proved to be the only communication channel for

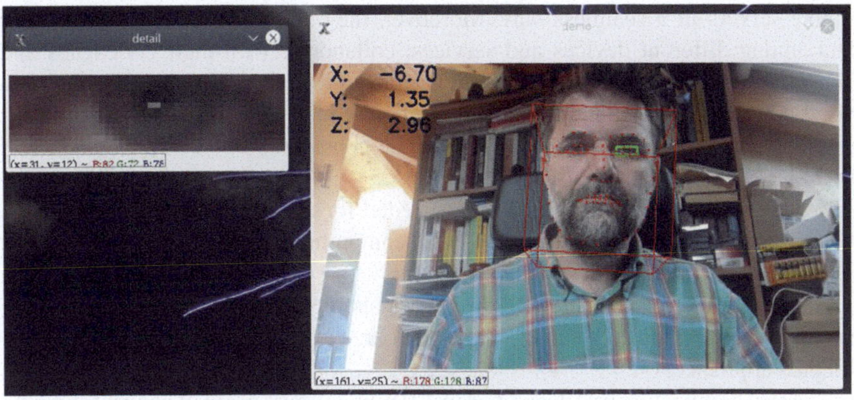

Fig. 2 Eye gaze detection and tracking from remote web camera

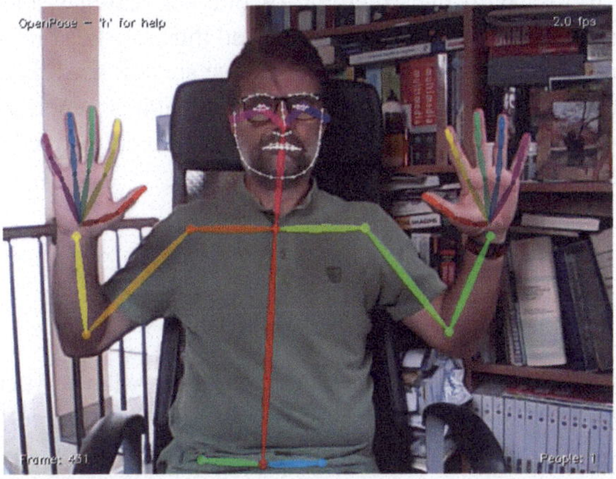

Fig. 3 Full-body pose recognition

some classes of impaired persons. However, most applications rely on specialized hardware;

- Full-body gesture and pose recognition (Fig. 3): several techniques are being tested with encouraging results.

5 Conclusions

Cloud-based platforms and architectures are currently the state of the art in Ambient Intelligence. In general, cloud services are useful and often necessary, as they shift the computational and storage burden from the users' devices to powerful and reliable remote servers in a seamless way. Moreover, they allow the users to share their data among different devices and services, collaborate with their colleagues and friends, and preserve the user data from accidental loss. However, the use of cloud-based services for user interaction and modelling may lead to severe issues and it is nowadays mainly motivated by commercial considerations, while the end users might well prefer other solutions.

We argue that cloud services are not needed for most smart home applications that do not search data on the Internet. Local computing is often faster, smarter, more secure and safe than cloud-based applications.

As an example, we illustrated how Red, a simple and cheap local architecture, can effectively provide the same services that are currently being offered by cloud-oriented appliances.

The Red architecture was developed in 2018, mainly as a proof of concept showing that local computing is currently able to support the development of effective and efficient smart and interactive agents, that interface both with the user and home automation platforms. With the exception of Internet searches, the Red platform does not rely on remote/cloud-based services, and thus satisfies our requisites.

Red is now more than one year old, however we are still actively developing and improving it. Hopefully, new applications and competing commercial products will soon emerge from the market, providing a convenient and effective local computing alternative to *clouds*.

References

Adrian, A.I., Ismet, P., Petru, P.: An overview of intelligent surveillance systems development. In: 2018 International Symposium on Electronics and Telecommunications (ISETC), Timisoara, pp. 1–6 (2018)

Ahonen, P., Alahuhta, P., Daskala, B., et al.: Threats and vulnerabilities. In: Safeguards in a World of Ambient Intelligence, The International Library of Ethics, Law and Technology, vol. 1. Springer Science+Business Media B.V., p. 143 (2010). ISBN 978–90–481–8786–7

Alexakis, G., Panagiotakis, S., Fragkakis, A., Markakis, E., Vassilakis, K.: Control of smart home operations using natural language processing, voice recognition and IoT technologies in a multi-tier architecture. Designs **3**, 32 (2019)

AlHammadi, A., AlZaabi, A., AlMarzooqi, B., AlNeyadi, S., AlHashmi, Z., Shatnawi, M.: Survey of IoT-Based Smart Home Approaches. In: 2019 Advances in Science and Engineering Technology International Conferences (ASET), Dubai, United Arab Emirates, pp. 1–6 (2019)

Amodei, D., Ananthanarayanan, S., Anubhai, R., et al.: Deep Speech 2: End-to-End Speech Recognition in English and Mandarin. In: Proceedings of The 33rd International Conference on Machine Learning, in PMLR, vol. 48, pp. 173–182 (2016)

André, E., Pelachaud, C.: Interacting with Embodied Conversational Agents, pp. 123–149. Springer US, Boston, MA (2010)

Battaglia, F., Iannizzotto, G., Lo Bello, L.: A person authentication system based on RFID tags and a cascade of face recognition algorithms. In: IEEE Transactions on Circuits and Systems for Video Technology, vol. 27, no. 8, pp. 1676–1690 (2017)

BBC Technology.: Amazon, Apple and Google face data complaints, BBC News. Online: https://www.bbc.com/news/technology-46944694 (2019)

Bengio, Y.: Learning deep architectures for AI. Found. Trends Mach. Learn. **2**(1), 1–127 (2009)
Bibri, S.E.: The current status, technological determinism, and unrealism of the AmI vision. In: The Shaping of Ambient Intelligence and the Internet of Things. Ambient and Pervasive Intelligence, vol. 10. Atlantis Press, Paris (2015)
Borah, J., Sarma, K.K., Gohain, P.J.: all pervasive surveillance techniques and AI-based applications: Current trends and challenges. In: Smart Devices, Applications, and Protocols for the IoT, IGI Global, pp. 54–82 (2019)
Cooper, C.: Windows 1.0: The flop that created an empire. CNET. CBS Interactive. Online: https://news.cnet.com/8301-10805_3-57613051-75/windows-1.0-the-flop-that-created-an-empire/ (2013)
Coskun, A., Kaner, G., Bostan, I.: Is smart home a necessity or a fantasy for the mainstream user? A study on users' expectations of smart household appliances. Int. J. Des. **12**, 1. Online: https://ijdesign.org/index.php/IJDesign/article/view/2938 (2018)
Elgan, M.: Why Amazon echo is the future of every home. Computerworld. online: https://www.computerworld.com/article/2844509/why-amazon-echo-is-the-future-of-every-home.html (2014)
Foster, M.E.: Enhancing Human–Computer Interaction with Embodied Conversational Agents, HCI (2007)
Fox, C.: Google and Facebook accused of breaking GDPR laws. BBC News. Online: https://www.bbc.com/news/technology-44252327 (2018)
Greenfield, A.: Everyware: The Dawning Age of Ubiquitous Computing, 1st edn, New Riders Publishing (2006)
ISTAG.: Scenarios for Ambient Intelligence in 2010. In: Ducatel, K., Bogdanowicz, M., Scapolo, F., Leijten, J., Burgelman, J.-C. (eds.) IPTS-ISTAG, Luxembourg, EC (2001)
Iannizzotto, G., Lo Bello, L., Nucita, A., Grasso, G.M.: A Vision and Speech Enabled, Customizable, Virtual Assistant for Smart Environments. In: 2018 11th International Conference on Human System Interaction (HSI), pp. 50–56 (2018)
Kim, K., Boelling, L., Haesler, S., Bailenson, J.N., Bruder, G., Welch, G.F.: Does a Digital Assistant Need a Body? The Influence of Visual Embodiment and Social Behavior on the Perception of Intelligent Virtual Agents in AR. In: Proceedings of the 17th IEEE International Symposium on Mixed and Augmented Reality (ISMAR 2018), Munich, Germany, 16–20 Oct 2018
Kramer, M., Yaghoubzadeh, R., Kopp, S., Pitsch, K.: A conversational virtual human as autonomous assistant for elderly and cognitively impaired users? Social acceptability and design considerations. In: Horbach, M. (ed.) INFORMATIK 2013. Informatik angepasst an Mensch, Organisation und Umwelt, Koblenz, Germany, vol. 220, pp. 1105–1119 (2013)
Lester, J.C., Converse, S.A., Kahler, S.E., Barlow, S.T., Stone, B.A, Bhogal, R.S.: The persona effect: Affective impact of animated pedagogical agents. In: Proceedings of the ACM SIGCHI Conference on Human Factors in Computing Systems, CHI '97, pp. 359–366, ACM, New York, NY, USA (1997)
Lorencik, D., Sincak, P.: Cloud robotics: Current trends and possible use as a service. In: IEEE 11th International Symposium on Applied Machine Intelligence and Informatics (SAMI), Herl'any, pp. 85–88 (2013)
Lynley, M.: Google unveils Google Assistant, a virtual assistant that's a big upgrade to Google Now. TechCrunch. AOL. Online: https://tcrn.ch/23Xl0pe (2016)
Mac-OS.: Apple Computer Inc. The Macintosh Product Introduction Plan. Stanford University Libraries & Academic Information Resources. typed ms, M1007, Apple Computer Inc. Papers, Series 7, Box 13, Folder 12, 7 Oct 1983
Norman, D.A.: The Invisible Computer. MIT Press, Cambridge, MA, USA (1998)
Prakash, E., Wood, J., Li, B., Clarke, M., Smith, G., Yates, K.: Games technology: Console architectures, game engines and invisible interaction. https://hdl.handle.net/2134/20295. Accessed 10 Nov 2019

Rogers, Y.: Moving on from Weiser's vision of Calm Computing: engaging ubicomp experiences. In: Dourish, P., Friday, A. (eds.) Proceedings of the 8th international conference on Ubiquitous Computing (UbiComp'06), Springer, Berlin, Heidelberg, pp 404–421 (2006)
Spiekermann, S., Pallas, F.: Technology paternalism—Wider implications of ubiquitous computing. Poiesis Prax. **4**, 6–18 (2006)
Streitz, N.A.: The disappearing computer initiative. Ercim News. Online: https://www.ercim.eu/publication/Ercim_News/enw46/streitz2.html (2001)
Streitz, N.A.: From Human–Computer Interaction to Human–Environment Interaction: Ambient Intelligence and the Disappearing Computer. In: Stephanidis, C., Pieper, M. (eds.) Universal Access in Ambient Intelligence Environments. Lecture Notes in Computer Science, vol. 4397. Springer, Berlin, Heidelberg (2007)
Velasco-Montero, D., Fernández-Berni, J., Carmona-Galán, R., Rodríguez-Vázquez, Á.: Performance analysis of real-time DNN inference on Raspberry Pi. In: Proceedings of SPIE 10670, Real-Time Image and Video Processing 2018, p. 106700F, 14 May 2018
Wadlow, T.: The Xerox Alto Computer. Byte Magazine. **6**(9) (1981)
Weiser, M.: The Computer for the 21st Century. Sci. Am. **265**, 94–104 (1991)
Weiser, M.: Ubiquitous computing. Computer **26**(10), 71–72 (1993)
Weiser, M., Brown, J. S.: Designing calm technology. POWERGRID J. (1996)
Weiss, B., Wechsung, I., Kühnel, C., Möller, S. Evaluating embodied conversational agents in multimodal interfaces. In: Computational Cognitive Science, vol. 1, p. 6 (2015)
Yaghoubzadeh, R., Kopp, S.: Toward a virtual assistant for vulnerable users: Designing careful interaction. In: Proceedings of the 1st Workshop on Speech and Multimodal Interaction in Assistive Environments, pp. 13–17. Association for Computational Linguistics (2012)
Yaghoubzadeh, R., Kramer, M., Pitsch, K., Kopp, S.: Virtual agents as daily assistants for elderly or cognitively impaired people. In: Aylett, R., Krenn, B., Pelachaud, C., Shimodaira, H. (eds.) Intelligent Virtual Agents, pp. 79–91, Springer, Berlin, Heidelberg (2013)
Yannakakis, G.N.: Game AI revisited. In: Proceedings of the 9th conference on Computing Frontiers (CF '12). ACM, New York, NY, USA, pp. 285–292 (2012)

Development of a Index for Sustainable Energy Technologies Based on an Intelligent Fuzzy Expert System

Fausto Cavallaro

Abstract The energy production activity can generate negative effects on the environment which must be taken into account. The traditional assessment models of environmental sustainability are in many cases affected by uncertainty. Fuzzy-sets have evidenced to be able to deal very well with uncertainty. In this paper an index based on an intelligent fuzzy inference system is proposed to assess the impact on the environment of the most important electricity power production technologies.

Keywords Fuzzy Inference System · Sustainabilty · Energy Technologies · Expert System

1 Introduction

Power plants activity can have a significant adverse impact on the environment increasing health risks and reducing the quality life of local communities. Then the environmental effects of electricity production processes can not be ignored. Many authors given important contributions about the relationship between energy production processes and ecological and human health impacts (Cavallaro and Ciraolo 2013, 2015; Shaaban and Scheffran 2017; del Río and Burguillo 2009; Troldborg et al. 2014; Mainali and Silveira 2015; Rovere et al. 2010; Onat and Bayar, 2010; Vera and Langlois 2007; Cavallaro 2015). But in many contexts the assessment models of environmental impact are affected by uncertainty. The uncertainty comes from the variability of the data and from imprecision that appears when observing or measuring the values of a variable (Cavallaro 2010). Fuzzy-set based approaches have evidenced to be able to deal very well with uncertainty in sustainability assessment.

In this paper we propose the development of an indicator based on fuzzy inference to assess the impact level on the environment of the energy technologies (see the Fig. 1). This paper is structured as follows: the Sect. 2 introduces the fuzzy inference

F. Cavallaro (✉)
Department of Economics, University of Molise, Campobasso, Italy
e-mail: cavallaro@unimol.it

D. Marino and M. A. Monaca (eds.), *Economic and Policy Implications of Artificial Intelligence*, Studies in Systems, Decision and Control 288,
https://doi.org/10.1007/978-3-030-45340-4_10

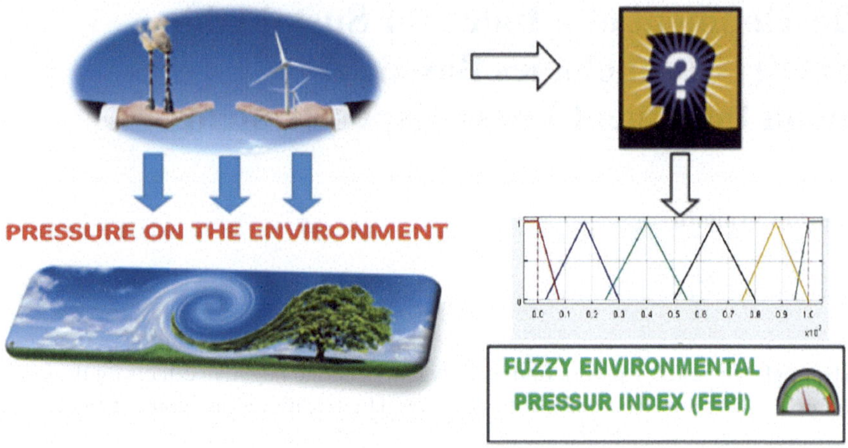

Fig. 1 The proposed sustainability index

theory, Sect. 3 shows the application of the proposed approach and the discussion, then in the Sect. 4 is reported the conclusion.

2 Fuzzy Inference System

In a fuzzy inference model (approximate reasoning) the reasoning process is based on a series of *if-then* rules as a kind of expert knowledge (Cornelissen et al. 2001; Öztaysi et al. 2013). The conditional statement (or proposition) contains a *premise*, the *if-part*, and a *conclusion*, the *then-part*. The knowledge incorporated in a fuzzy control system is made up of a collection of several parallel rules in the form '*if X is A then Y is B*', or more generally '*if* X_1 *is* A_1 *and …and* X_n *is* A_n *then Y is B*', where A, A_n, B are fuzzy sets (Dubois et al. 2007). The *knowledge base*, which contains the general knowledge concerning a problem domain, and which connects antecedents with consequences, premises with conclusions, or conditions with actions. A fuzzy inference system (FIS) is composed of three blocks (Saberi Nasr et al. 2012). The first one, *fuzzification*, converts the crisp value input to a linguistic variable using the membership functions kept in the knowledge base. To the second block, the *inference engine*, is assigned the task of evaluating the input's degree of membership to the fuzzy output sets using the fuzzy rules. Finally, the defuzzifier block transforms the fuzzy output into a crisp value (Fig. 2).

The inference engine is the heart of the FIS that reproduce the human approximate reasoning. The inference stage utilizes the fuzzy input values to activate the inference rules and generate the fuzzy output value. The popular approaches to fuzzy inference are Mamdani type and the Takagi-Sugeno-Kang type (Klir and Yuan 1995). An

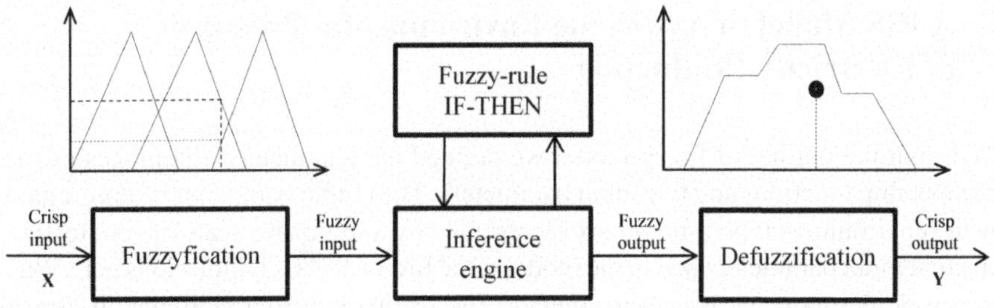

Fig. 2 Fuzzy Inference System

illustration of Mandami FIS is shown in Fig. 3. The method most frequently used for the composition of fuzzy relations is the max-min technique.

For example, a two-rule max-min composition Mamdani FIS model is the following (Saberi Nasr et al. 2012):

$$\mu_{Ck}(Z) = max[min[\mu_{Ak}(input(x)), \mu_{Bk}(input(y)]] \quad k = 1, 2 \ldots r \tag{1}$$

where μ_{Ck}, μ_{Ak} and μ_{Bk} are the membership functions of output *Z* for rule *k* input *A* and input *B*. Even so, other minor fuzzy inference methods have been proposed.

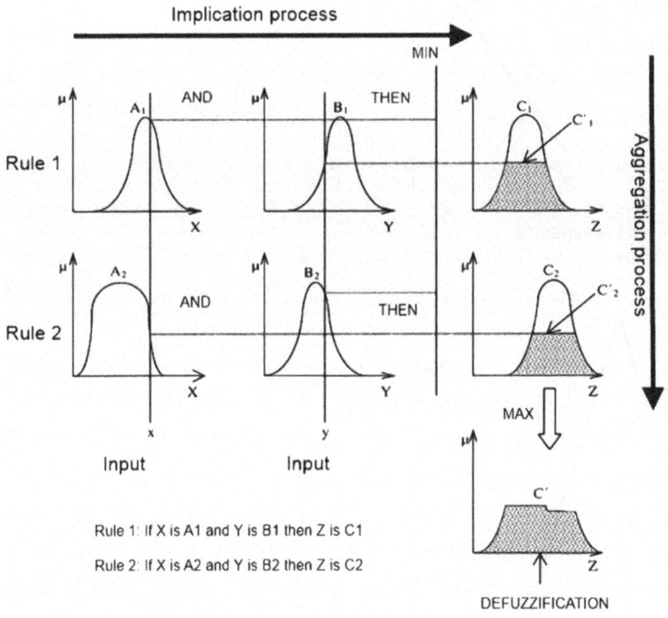

Fig. 3 Mamdani FIS

3 A FIS Model to Assess the Environmental Pressure of Electricity Production

To design the outline of fuzzy model we defined the parameters, the ranges of the membership functions and five input parameters. The output parameter is represented by the environmental pressure index represented by a score on a scale of zero to 100. The first input parameter used in the model is the life-cycle "CO_2 emissions per kWh" of electricity produced by energy sources. The second indicator employed is "Land use" that represents the area used for power generation. The third indicator is "Social acceptance" that refers to the level of acceptance by the local population of the power plants projects. The four indicator corresponds to the "Impact on ecosystem". It is connected to every environmental change can have ecological effects. Finally, the last indicator defines the "Visual intrusion". This concerns the visual inconvenience that can cause a power plant in a specific area. The landscape of different sites and the possibility to integrate the plant in the surroundings should be considered.

The input parameters are classified on the base of three membership functions such as *low, medium, high*. In the Figs. 4 and 5 are reported respectively the membership

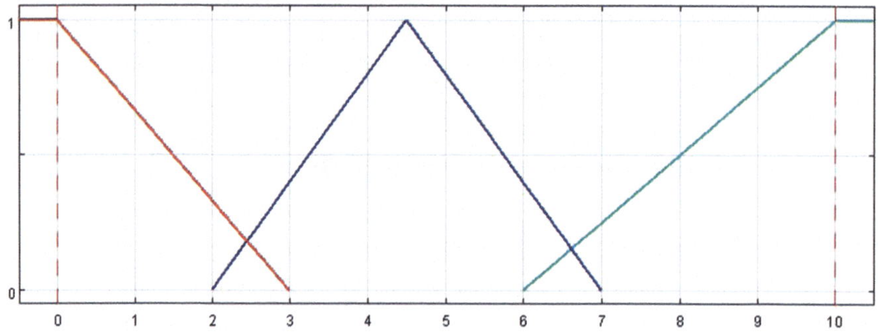

Fig. 4 MS social acceptance

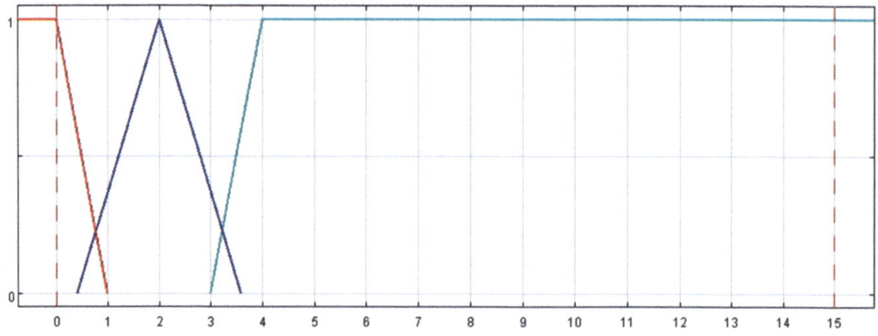

Fig. 5 MS Land use

Table 1 Input parameters

Input variables	Linguistic values	Base variable	Range
CO_2 emissions	(Low, middle, high)	gCO_{2eq}/kWh	[0–1000]
Land use	(Low, middle, high)	km^2/TWh	[0–15]
Social acceptance	(Low, middle, high)	score	[0–10]
Impact on ecosystem	(Low, middle, high)	score	[0–1]
Visual intrusion	(Low, middle, high)	score	[0–10]

functions of the "Social acceptance" and "Land use". The linguistic variables are low (red line) medium (blue line) and high (cyan line) represented respectively for the first one parameter by the Eqs. (2), (3) and (4) while for the second parameter by the Eqs. (5), (6) and (7). In the Table 1 are reported the linguistic values and the variables of the selected input parameters.

$$\mu_{low}(x) = \left\{ \begin{matrix} 0 & x \geq 3 \\ \frac{3-x}{3} & 0 \leq x < 3 \end{matrix} \right\} \tag{2}$$

$$\mu_{medium}(x) \left\{ \begin{matrix} \frac{x-2}{4.5-2} & 2 \leq x < 4.5 \\ 1 & x = 4.5 \\ \frac{7-x}{7-4.5} & 4.5 \leq x < 7 \end{matrix} \right\} \tag{3}$$

$$\mu_{high}(x) = \left\{ \begin{matrix} 0 & x \leq 6 \\ \frac{x-6}{10-6} & 6 \leq x < 10 \end{matrix} \right\} \tag{4}$$

$$\mu_{low}(x) = \left\{ \begin{matrix} 0 & x \geq 1 \\ 1-x & 0 \leq x < 1 \end{matrix} \right\} \tag{5}$$

$$\mu_{medium}(x) \left\{ \begin{matrix} \frac{x-0.5}{3.5-0.5} & 0.5 \leq x < 3.5 \\ 1 & x = 2 \\ \frac{3.5-x}{3.5-2} & 2 \leq x < 3.5 \end{matrix} \right\} \tag{6}$$

$$\mu_{high}(x) = \left\{ \begin{matrix} 0 & x \leq 3 \\ \frac{x-3}{4-3} & 3 \leq x < 4 \\ 1 & x \geq 4 \end{matrix} \right\} \tag{7}$$

Once the calculations are made after defuzzification step we obtain as result, a value (on a scale of zero to 100) that represents the Fuzzy Environmental Pressure Index (FEPI). This value is a measurement of the environmental pressure caused by energy technologies. The linguistic ratings of the FEPI are described as *very low*, *low*,

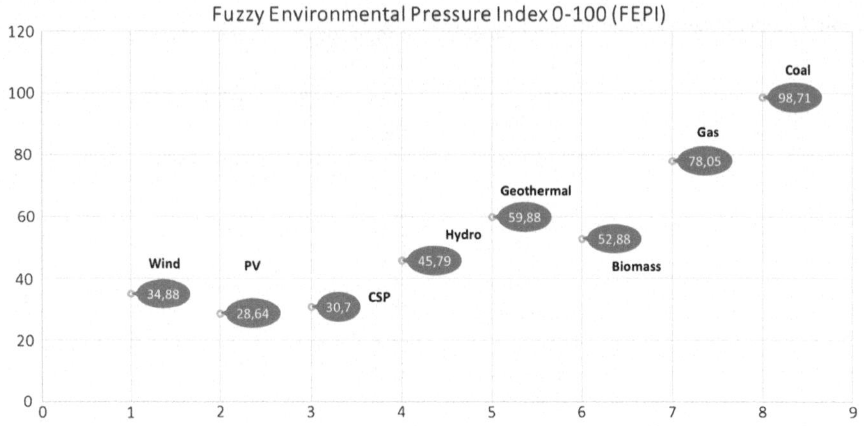

Fig. 6 FEPI results

middle, *middle high*, *high*, and *very high*. We tested the model for the main energy technologies and we obtained the following values of FEPI (Fig. 6): Wind (34.88), PV (28.64), CSP (30.70), Hydro (45.79), Biomass (52.88), Geothermal (59.88) and finally Gas (78.05) and Coal (98.71).

4 Conclusions

In this paper, we proposed an intelligent fuzzy expert system to assess the environmental sustainability of energy production technologies. An FIS contains information and experience of an expert in the design of a system that controls a process whose input–output relations are defined by a set of fuzzy control rules. The use of traditional mathematical approaches to dealing with ill-defined problems in many cases is not suitable. Instead an intelligent machine learning that employs fuzzy *if-then* rules can model the uncertainty system such as the environmental sustainability. An index based on fuzzy inference system has been proposed to measure the pressure from energy technologies on the environment. Testing phase produced encouraging results then the future research will deal with the improvement of the model.

References

Cavallaro, F.: Fuzzy TOPSIS approach for assessing thermal-energy storage in concentrated solar power (CSP) systems. Appl. Energy **87**(2), 496–503 (2010)

Cavallaro, F.: A Takagi-Sugeno fuzzy inference system for developing a sustainability index of biomass. Sustainability **7**, 12359–12371 (2015)

Cavallaro, F., Ciraolo, L.: Sustainability assessment of solar technologies based on linguistic information. In: Cavallaro, F. (ed.) Assessment and Simulation Tools for Sustainable Energy Systems: Theory and Applications, Springer, Berlin (2013)

Cavallaro, F., Ciraolo, L.: Design and implementation of a fuzzy inference model for mapping the sustainability of energy crops. In: Garcia Cascales, M.S., Sanchez Lozano, J.M., Masegosa Arredondo, A.D., Cruz Corona, C. (eds.) Soft Computing Applications for Renewable Energy and Energy Efficiency, IGI Global (2015)

Cornelissen, A.M.G., Van den Berg, J., Koops, W.J., et al.: Assessment of the contribution of sustainability indicators to sustainable development: a novel approach using fuzzy set theory. Agr. Ecosyst. Environ. **86**(2), 173–185 (2001)

del Río, P., Burguillo, M.: An empirical analysis of the impact of renewable energy deployment on local sustainability. Renew. Sustain. Energy Rev. **13**, 1314–1325 (2009)

Dubois, D., Esteva, F., Godo, L. et al.: Fuzzy-set based logics—an history-oriented presentation of their main developments. In: Gabbay, D.M., Woods, J. (eds) Handbook of the History of Logic. vol. 8, Elsevier BV (2007)

Klir, G.J., Yuan, B.: Fuzzy Sets and Fuzzy Logic—Theory and Applications. Prentice Hall, New Jersey (1995)

Mainali, B., Silveira, S.: Using a sustainability index to assess energy technologies for rural electrification. Renew. Sustain. Energy Rev. **41**, 1351–1365 (2015)

Onat, N., Bayar, H.: The sustainability indicators of power production systems. Renew. Sustain. Energy Rev. **14**, 3108–3115 (2010)

Öztaysi, B., Behret, H., Kabak, Ö., et al.: Fuzzy inference systems for disaster response. In: Vitoriano, B., Montero, L., Ruan, D. (eds.) Decision Aid Models for Disaster Management and Emergencies, Atlantis-Springer (2013).

Rovere, E.L.L., Soares, J.B., Oliveira, L.B., Lauria, T.: Sustainable expansion of electricity sector: sustainability indicators as an instrument to support decision making. Renew. Sustain. Energy Rev. **14**, 422–429 (2010)

Saberi Nasr, A., Rezaei, M., Dashti Barmaki, M.: Analysis of groundwater quality using Mamdani Fuzzy Inference System (MFIS) in Yazd Province. Iran. Int. J. Comput. Appl. **59**(7), 45–53 (2012)

Shaaban, M., Scheffran, J.: Selection of sustainable development indicators for the assessment of electricity production in Egypt. Sustain. Energy Technol. Assess. **22**, 65–73 (2017)

Troldborg, M., Heslop, S., Hough, R.L.: Assessing the sustainability of renewable energy technologies using multi-criteria analysis: suitability of approach for national-scale assessments and associated uncertainties. Renew. Sustain. Energy Rev. **39**, 1173–1184 (2014)

Vera, I., Langlois, L.: Energy indicators for sustainable development. Energy **32**, 875–882 (2007)

Blockchain and Charities: A Systemic Opportunity to Create Social Value

Ardian Foti and Domenico Marino

Abstract Countless media reports have exposed scandals concerning charities and non-profit organisations, involving mismanagement of funds, or even outright theft and misappropriation of funds earmarked for social works and projects. The aim of this paper is to show how Blockchain technology can be reconciled with the operating mechanisms of charities.

Keywords Blockchain · Charity · Volunteering

1 Introduction

The aim of this paper is to show how Blockchain technology can be reconciled with the operating mechanisms of charities.[1] To 'do' charity, by donating money or volunteering, one should also have reasonable confidence that the charitable organisation will spend appropriately and effectively the resources donated to it by individual citizens and donor organisations, such as governments, companies and foundations. Moreover, this confidence, being based on a relationship of trust concerns not only spending and project delivery capacity but also the full transparency of charity project management processes. Countless media reports have exposed scandals concerning charities and non-profit organisations, involving mismanagement of funds, or even outright theft and misappropriation of funds earmarked for social works and projects.[2]

[1] https://www.binance.vision/blockchain/blockchain-use-cases-charity.

[2] https://europa.today.it/attualita/si-allarga-lo-scandalo-dei-cooperanti-del-sesso-120-casi-in-ong-britanniche-anche-per-pedofilia.html.

A. Foti
Strategie Soluzioni Servizi, Rome, Italy
e-mail: ardianfoti@gmail.com

D. Marino (✉)
Mediterranea University of Reggio Calabria, 89100 Reggio Calabria, Italy
e-mail: dmarino@unirc.it

D. Marino and M. A. Monaca (eds.), *Economic and Policy Implications of Artificial Intelligence*, Studies in Systems, Decision and Control 288,
https://doi.org/10.1007/978-3-030-45340-4_11

2 The Charity Sector

If we extend the analysis to NGOs and international philanthropic foundations, we can get to the heart of the economic and management matters that are affected by the greatest administrative and project design problems. Here we can experience directly the disproportion between actual costs and various types of financial fraud.[3] The utter lack of transparency is on the one hand compensated by the certification of the financial statements by auditing firms, which in a few cases operate in conflict of interest, even though this instrument is applied by foundations and NGOs with transnational presence.[4] The European Court of Auditors has developed a protocol for performing financial compliance audits and performance audits with all the parties concerned, as well as guidelines for detecting fraud in cost increases and in funding to third parties involved in projects.

Nevertheless, national governments need accounting and financial clarity about the funding sources of NGOs and their statutory aims, for purposes of national security.[5] As is well known, all too often purported religious organisations are actually fronts for terrorist networks and financial transactions that have no charitable purpose, but support families and groups of extremists pursuing the interests of countries and religious foundations contrary to the democratic order of Western countries.[6] This has caused a significant decline in donations, especially from young people; in some cases NGO funding policies in Palestine and in other areas with high levels of ethnic and social conflict have come under scrutiny. The establishment of NGO Monitor has created quite a few problems for various philanthropic entities and NGOs that 'unwittingly' finance terrorist networks and systems engaged in international fraud or in exerting political influence on national governments. Some NGOs may even support anti-Semitic associations, ethnic propaganda or infiltration of covert economic agents for purposes that have little to do with charity.[7] The founders, financial flows and donors of NGOs and countless uncertified local associations have been subjected to in-depth monitoring, revealing that all too often the activities of charities, ostensibly aimed at promoting inclusive policies, actually serve as a cover for quite different purposes.[8]

---

[3]https://www.theguardian.com/global-development-professionals-network/2016/aug/25/fraud-aid-agencies-upset-public.

[4]https://www.eca.europa.eu/Lists/ECADocuments/SR18_35/SR_NGO_FUNDING_EN.pdf.

[5]https://reliefweb.int/report/world/counter-terrorist-financing-and-humanitarian-security.

[6]https://www.gatestoneinstitute.org/11643/ngo-industry-terrorism.

[7]https://www.ngo-monitor.org/.

[8]https://www.analisidifesa.it/2019/03/terrorismo-criminalita-e-contrabbando-gli-affari-dei-jihadisti-nellultimo-rapporto-icsa/.

3 Blockchain Technology

Blockchain technology is a new tool whose use is rapidly expanding in finance and in other everyday applications. A blockchain is basically a certified set of data records, which cannot be changed, which is managed by a cluster of computers that do not belong to a single user. Each of the data blocks is protected and linked to each other by means of cryptographic principles (chain). In order to modify one element, it is necessary to have access to all the computers that make up the chain in order to be able to decrypt the data.

Paradoxically, blockchain is often misconstrued by poorly informed mainstream media as a technological tool antithetical to the principles of transparency and accountability. Nothing could be farther from the truth. This misconception stems from the non-transparent use of cryptocurrencies by traders. Since they are based on blockchain technology, cryptocurrency exchanges and exchanges on token platforms are often assumed to be associated with money laundering and international criminal networks. This is simply not the case, as is shown by a more in-depth look at how cryptocurrencies are exchanged. The process is very lengthy and requires complex IT procedures for financial and banking triangulation; it is monitored by governments and IT security companies dealing with digital identity theft and cryptocurrency portfolios. This general information about blockchain technology is widely available from media outlets and scientific publications on the internet. Moreover, building value through mining is becoming less and less cost-effective unless one uses technological tools and servers able to cut the costs of electricity and of cooling the machines. The rumours that tend to overestimate the risks and threats of new blockchain technologies are to be considered fake news. The ductility of this technology lies in the certification of nodes and codes.

This property allows an extremely wide use of this tool, which can be applied to a number of sectors. Blockchain technology can help to improve a large number of processes and activities, making them more transparent, safe and efficient. For example, it can be used in the fields of banking, insurance, contracts, education, e-voting, cybersecurity, health and transport.

4 How to Integrate Charities and Blockchain

How can charities benefit from blockchain technology? This technology offers a number of features that can successfully be exploited by charitable organisations:

- Certification of and for donors
- Certification of services and research
- Financial certification and transparency
- Validation of networking and crowdfunding
- Accountability according to national and international standards
- Security of transactions

- Big data management in accordance with the EU's GDPR
- Computer protocols registered with the competent agencies
- Environmental and social sustainability for companies
- Integrated database of projects and scientific research.

Blockchain technology can also offer significant advantages to non-profit banking organisations and to trade unions, associations and all those non-profit organisations whose operations combine public and private resources to achieve their statutory purposes. Controls by the competent national authorities would also be carried out in full compliance with existing personal data protection legislation and with the fiscal parameters necessary for companies and independent donors. Even the financing of political parties and political movements could be managed entirely in blockchain. This implies the legislator's full political commitment to transparency in the management of funding and political activities.

5 Final Remarks

There should be no more excuses for postponing the use of advanced technology in our societies. We need an ethical and moral revolution that fully addresses our conscience and knowledge, to protect the world and human societies built on ideas and principles which have become non-negotiable. The economic principles of sustainability and scarcity can be fully implemented via transparency, the efficient use of resources and the systemic capacity to combine social activities with the highest philanthropic principles which are embraced not only by religions but also by secular societies. If there is a principle of syncretism between thoughts, ideas and religions we owe it to blockchain technology, to its potential to match effectively with artificial intelligence and the countless functionalities that will be offered by the evolution of cybernetic systems. Blockchain is a universal tool that can change the world and social relations. What we need is greater awareness and an ethical and responsible adaptation of the role we intend to play in building our future common home.

AI and Teaching Approach in High School

Marino Tommaso and Pecchio Paola

Abstract We discuss the importance of Information and Communication Technologies (ICT) in High School education. The Artificial Intelligence role is analysed not only for Informatics lessons but as a global approach in interdisciplinary school projects. The processes of the AI training and the students training are compared in order to encourage students to have awareness for their knowledge acquisition. Moreover AI applications can offer an important technical and methodological support for students with special needs, in order to facilitate their inclusion into ordinary classes.

Keywords Artificial Intelligence · Education · Special educational needs

1 Introduction

Many benefits of AI Technologies in educational practice are widely known (Drigas and Ioannidou 2012). For example, it helps in automatic marking, allowing greater objectivity in the evaluation of some kinds of tests that require a simple answer. Moreover, teaching software can be customized for each student, permitting welcome support for students with SEN (Special Educational Needs) in large classes.

AI applications can also become personalized topics for each student. Their use can help students to follow a path of in-depth study especially where they are lacking in the education plan. Their use can open an interesting interdisciplinary perspective. Additionally, AI applications can realize a personalized tutor, so that each student can optimize his study method. They also provide ease in collecting feedback of learning,

M. Tommaso (✉)
IIS CURIE LEVI, Turin, Italy
e-mail: marino@liceomcurie.it

AIF (Association for Physics Teaching), Turin, Italy

World Space Week, Torino, IT, Italy

P. Paola
Istituto Scolastico Sacra Famiglia, Liceo Scientifico, Turin, Italy

D. Marino and M. A. Monaca (eds.), *Economic and Policy Implications of Artificial Intelligence*, Studies in Systems, Decision and Control 288,
https://doi.org/10.1007/978-3-030-45340-4_12

giving an important aid to self-assessment. They change the teacher role : instead of being the unique person in charge of the learning process, the teacher becomes a tutor and the real protagonist becomes the student, by making the learning system more "friendly" and less "intimidating", by changing the way we teach and learn. Furthermore they change how information is retrieved and they can also help students to find information by using different platforms and not just one search engine.

Finally, many papers underline the role of AI systems in supporting students with learning difficulties and special needs (Drigas and Ioannidou 2012). Among several ways to overcome difficulties in the learning process of students with Special Educational Needs (SEN), AI tools can be one of them due to many recognized benefits. AI is generally connected to intelligent artificial agents which are able to analyze the environment, in order to act in a productive way (Russell and Norvig 2003). See also a family with an autistic child that through the use of AI technology has been able to achieve good learning results that are difficult to achieve with traditional methods (cft "A Siri with love, Judith Newman).

Despite the important advantages cited above, educational practice and lesson procedures tend to lack use of AI technologies.

In this paper a proposal of insertion of special projects in lesson time with high school students and a draft method to support learners with SEN by using AI is discussed.

2 Intersections Between AI and Students

In the Italian school, teaching takes place on three distinct levels: knowledge, abilities and skills. On the attached questionnaire, students were asked to compare the learning methods of human beings (in particular their own method) and neural networks, in order to introduce a further consideration: Are knowledge and skills sufficient to grow as a citizen? They realized that the achievement of skills would be useful later at university and/or at work but it represents a challenge rather than a goal. A high level of personalization in teaching and taking into account the resources, human and financial, of the school system became crucial and really often decisive for each student.

Each person has his own way and "decoding time" of the information taught. This is the basic characteristic of the specific cognitive structure. Unfortunately the lesson time proposed by the teacher is hardly the best one for all students and in particular for students with SEN. In this situations, the use of AI could support students in decoding information, speeding up the process of writing notes with the consequent identification of concepts and keywords.

Moreover the Italian school system has long been endowed with a digital plan, a document of the Ministry of Education, University and Research, that strongly encourages an overall innovation strategy of the Italian school and for a new positioning of its educational system in the digital age (cit.). In order to check if students are really interested in this kind of procedure and wheter most of them would succeed in these more technological lessons, our statistical survey was made in order to analyze if our students were aware of the presence of AI in their own life. It was also checked if they would like to introduce the use of AI systems in lessons. They are generally expected to be only partially aware of the daily use of A.I. systems. A sample of 48 students was tested on their familiarity with AI systems, in order to collect useful information on definition and use of AI in their daily life as high school students. They could give an answer even if they did not know everything about the meanings of words and topics related to AI systems (Data mining, deep learning, machine learning). Our survey was carried out on a homogeneous sample (aged 14 to 18) and tried to represent only a first cognitive step that deserves further investigation.

It turned out that they are generally aware of the presence of AI systems in many devices (Fig. 1) and perform some functions carried out so far exclusively by humans (reservation of restaurant, place a call for someone else or create an artificial profile using someone data) (Fig. 2). A large majority (77% use AI in their life) believe that AI systems will be commonly used and an even greater amount (83.3% positive vs. negative use) believes that the use of AI is a positive factor for improving quality of life.

These first results show the importance that AI systems hold in people's lives. They must be implemented and studied also in classrooms. Students must know these systems from the first years of school and not only as users. In the complete report additional elements to introduce the study of AI systems in schools were included.

These attachments report the full questionnaire.

Therefore some systems based on AI technologies were proposed as a subject of a trial in order to improve their learning strategies: Amazon echo (based on the Alexa system), a Vector robot for educational uses, Spotify, ability to recommend songs based on the tastes of the user, Netflix which suggests the series starting from the user's tastes and the Google world which provides articles, news and indications

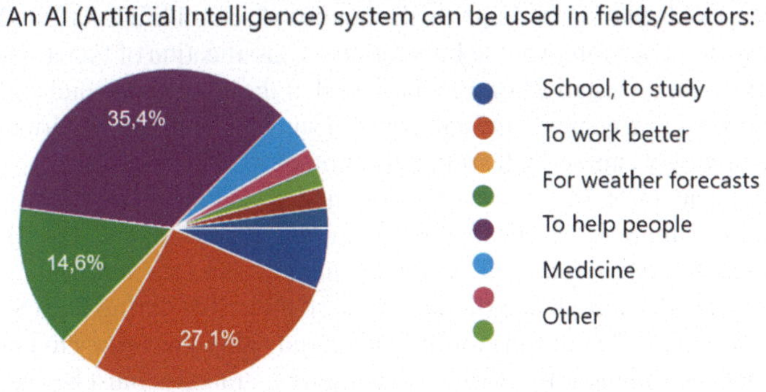

Fig. 1 Responses on test: how AI systems can be used in different fields

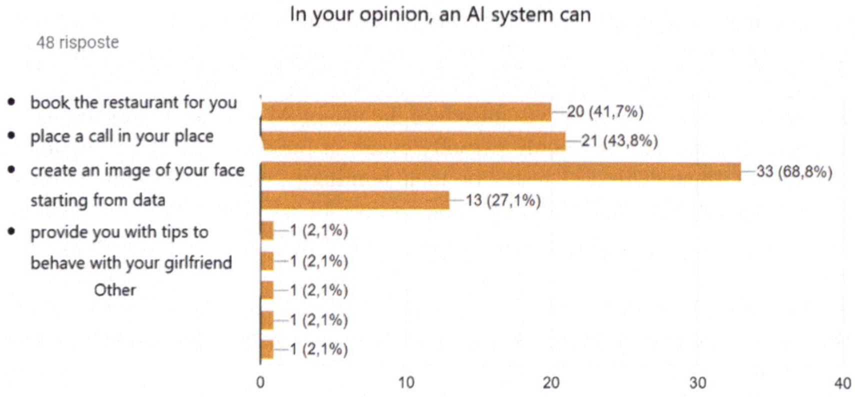

Fig. 2 Responses on test: what AI system can instead of students

of our cultural interest with a geographical base. A further study is scheduled in order to continue the comparison between AI systems learning and students learning procedures. Students will be involved in the investigation of the process of their knowledge acquisition.

3 AI and Special Needs Education

Smartphone applications can be considered in order to promote the scholarly success and guarantee an optimal education as well as the emotional and relational growth for all students. Many smartphone apps works thanks to A.I. structure. Besides the fundamental step of an early and correct diagnosis, the introduction of compensatory tools that ensure a systematic and uninterrupted learning process at school and at home is considered a human right (Italian Law 2010). Italian scholar integration underlines the inclusive approach and requires a strong and incisive use of supporting strategies and tools in order to improve the quality of school life for all students, not only those with SEN. Moreover the International Classification of Functioning (ICF) of the World Health Organization Emphasises that the functioning and ability of an individual occurs in a context, and can concern students in different periods of their school life, not only caused by learning disabilities.

Despite the attention of the scientific community and the creation of specific AI environments to improve the life of people with special educational needs (Lanzilotti and Roselli 2007), AI tools are still relatively unknown and not very widely diffused as school praxis. Moreover some students realized their SEN during High School and later the choice of a different and more effective study plan. According to Italian Ministerial directives (Italian Law 2013), compensatory strategies must be available for these students. Smarthphones and A.I. applications were used to support systematically students in our project. The smartphone is the most used and diffuse electronic instrument among the High School generation but actually not the most conveniently used by teenagers. A standard procedure was proposed in a High School class where students with no strong SEN represent the approximatively 15% of the students of each class involved. Smartphone applications were considered in order to allow each student with or without SEN to use the same tools. This way everybody discovered his favourite way to study during Mathematics lessons.

First, the use of smarthphone as learning support was proposed and discussed during lessons. By analyzing a standard lesson of Mathematics on a specific topic, students and the teacher tried to build a standard inclusive lesson strategy. A cooperative approach between students and teachers allowed encouraging the learning process by checking and integrating different strategies.

Second, some smarthphone free-applications were used in order to support lessons with AI technologies. Apps that allow automated transcription service were used in order to reproduce the crucial few minutes of the explanation. Vocal assistant applications were used to search definitions and examples on the web. The PICAI application was used to optimize photos of exercises and examples written on a blackboard.

Third, lesson materials were checked by the teacher and uploaded on the virtual classroom.

This experimental procedure evidenced many positive aspects. Students could realized a new and technological way of learning at school. Many of them found it more interesting and effective than standard lessons. They also learned the importance of a structured way of work. Moreover, cooperative and technological learning revealed that everyone could give a personal contribution to common work and each could also gain an advantage in his own personal educational program.

4 Conclusions

The aim of this study intends to continue the investigation of the role of AI systems in education. Even if many papers have been proposed on this topic, there are still many processes still to investigate, especially if the students would be authors of the study with teachers.

The results of an introductive analysis of intersection between students and AI systems is reported here. Also students with SEN were involved in order to analyze the success of their integration in class.

Certainly the use of these systems in teaching procedures is a complex subject and this article would like to provide a brief contribution to related issues that involves those of AI systems in situations with Handicap and SEN as well as that concerning standard processes of learning.

Acknowledgements The authors are very grateful to all reviewers (Prof. Albert Werbrouck and Prof. Giuseppina Rinaudo Dept of Physiscs—University of Turin) for their valuable suggestions and comments to improve the quality of the manuscript.

References

Drigas A.S., Ioannidou R.E.: Artificial intelligence in special education: a decade review. Int. J. Eng. Educ. **28**(6), 1366–1372 (2012) (Printed in Great Britain)

Guilherme, A.: AI and education: the importance of teacher and student relations. AI Soc. **34**(1), 47–54 (2019)

Italian Law, Legge 170/2010, Nuove norme in materia di disturbi specifici di apprendimento in ambito scolastico

Italian Law, Circolare Ministeriale n.8/2013, Prot. 561

Lanzilotti, R., Roselli, T.: An experimental evaluation of Logiocando, an intelligent tutoring hypermedia system. Int. J. Artif. Intell. Educ. **17**, 41–56 (2007)

Russell, S.J., Norvig, P.: Artificial Intelligence: A Modern Approach, 2nd edn. New Jersey (2003)

Artificial Intelligence and Protection of National Interest. Challenges and Opportunities for Economic Intelligence

Pietro Stilo

Abstract That of national interest, is a very broad concept that certainly concerns a collective interest which by its nature is distinguished from private or individual interest and therefore for this motivation it must be protected and guaranteed as the interest of a given community that insists on a limited space, the reading that we want to offer in this sense is very geopolitical as well as geo-economics, not a political model I study but instead a model that focuses on three key points: interest, space and time. An important aspect whorty of analysis is the economic security, as a national interest like other aspects of national security. We therefore begin to discuss the economic issue which of the national interest to be covered also using the Intelligence tool, as a fundamental sector of the State for the protection of national interests.

Keywords Economic intelligence · Artificial intelligence · National interest

At the end of the last century Alessandro Corneli wrote a very interesting article in the 7th number of the Intelligence Magazine "Gnosis" entitled "The Intelligence Services and the National Interest" in his piece Corneli highlights.

> For several decades, the idea of the state as a "sovereign entity", which does not recognize any other authority above and outside itself, it must deal with the reality of an increasingly dense network of international (and in some cases supranational) agreements that have largely reduced or cancelled it, going beyond the phase of pure, and always existed, interdependence. To this must be added that the liberalization processes (of people's movements, of ideas, of capital, of goods, of services, of knowledge, of information) have increased the number of national subjects, understood as operating within the framework of the nation-state, increasingly operating and thinking in transnational terms, which is something different from international. If military security, understood as the set of conditions that allow the normal and orderly survival of the state, continues to rely on conceptually tested parameters, even if continuously updated to the various forms of threats, the same cannot be said for economic security of a state, understood as the set of conditions that allow the growing satisfaction of elementary and secondary needs, containing those threats that endanger it through the fall in production and trade, the high level of unemployment, the significant incidence of illegal

P. Stilo (✉)
Mediterranea University of Reggio Calabria, 89100 Reggio Calabria, Italy
e-mail: pietro.stilo@unirc.it

D. Marino and M. A. Monaca (eds.), *Economic and Policy Implications of Artificial Intelligence*, Studies in Systems, Decision and Control 288,
https://doi.org/10.1007/978-3-030-45340-4_13

> forms of economic activity, the hostile action of other subjects towards money or productive sectors. As for economic security, the analysis must consider the increasingly meaningful structure that is summarized in the concept of globalization, which changes the conceptual frameworks and operational criteria of the nation-state. The pursuit of military security and economic security, with which political security can be associated, understood as maintaining the conditions for exercising democratic freedoms, is the object of national interest. It is therefore appropriate to proceed step by step: the **first** is to examine the different traditional interpretations of national interest; the **second** is to analyse the current Italian regulatory framework; the **third** is to consider the changes taking place to try to identify the possible regulatory and operational paths that can guide the pursuit of the national interest and insert the role of the Intelligence Services in them. (https://gnosis.aisi.gov.it/sito/Rivista7.nsf/ServNavig/2)

In this article he highlights an important aspect and at the time probably still not relevant, namely economic security, as a national interest like other aspects of national security and therefore of national interest, such as military security, and so on. We therefore begin to discuss the economic issue which of the national interest to be covered also using the Intelligence tool, as a fundamental sector of the State for the protection of national interests.

That of national interest, is a very broad concept that certainly concerns a collective interest which by its nature is distinguished from private or individual interest and therefore for this motivation it must be protected and guaranteed as the interest of a given community that insists on a limited space, the reading that we want to offer in this sense is very geopolitical as well as geo-economics, not a political model I study but instead a model that focuses on three key points: interest, space and time. So from what can be deduced from this centre of analysis, the national interest thus understood must certainly be taken into consideration, analysed and studied from a cultural and certainly not ideological point of view (at least here), since if it is It is true that in some cases in political language it is taken into consideration in moments of national unity in the face of a need or a challenge, or during election campaigns to attract the electorate often of one political party more than another. In legal language, however, it has several facets: historical, cultural, geopolitical and geo-economics, aspects that we try to highlight here and that have experienced a strengthening with the globalization process, and in particular in the aftermath of the fall of the Berlin Wall, "limes" dividing a post-World War II world idea and a new post-Cold War phase, a phase in which the world has experienced intense and rapid changes like never before, with a loosening of national borders and a consequent strengthening of local and therefore national identities necessary to keep the interests of the national koiné clear and outlined and therefore the idea of protecting the national interest has been strengthened (almost like an antibody) being the limits and boundaries guaranteed by the equilibrium of that system that having collapsed, had also brought behind it an easing of national borders and an intensification of new challenges and threats for the national communities, more liquid and interdependent between them, stimulated and driven by drivers such as new technologies in the field of communications and transport, which have accelerated a process that has seen new players and new opportunities appear on the global scene, but also new challenges never considered before, since they are covered by the protective umbrella of the two opposing blocks.

In this context it is necessary, in order to better analyse the protection of the national interest, also in the light of the new developments dictated by technological innovation especially in the field of artificial intelligence, IOT and new communication processes and in industry, to face the intelligence sector, in particular the economic one, as a fundamental tool for the protection of the state as a whole. Fundamental sectors to better face the challenges and opportunities of today's globalized world, a world that has profoundly changed compared to the past and is constantly evolving, in which the role of Intelligence has become increasingly important and essential. A world that has radically changed since the fall of the Berlin Wall onwards (as mentioned above), all this has brought unusual challenges, but also new opportunities for agents in the field, innovations that are often attributed to changes in globalization, encouraged by the development of modern information and communication technologies and driven by the loosening of national borders. So also the security and protection of national interests has encountered changes with new threats that have arisen, increasingly multidimensional and global, therefore the military dimension is no longer the subject of attention of the intelligence services, but also other heterogeneous challenges and liquid and precisely for this reason difficult to contain by a physical barrier, among them for example: climate change, pandemics, mass migration, economic and financial instability, transnational organized crime, international terrorism, cybercrime.

But what is Intelligence? It is a cycle of analysing information and data. Starting from this concept, we can immediately understand that the first element on which we must focus our attention is information and data, but what information and what data should be sought? Surely this is an input that starts from the decision maker, who must establish what information and what data he intends to obtain through Intelligence, for the objectives he aims to achieve. The word Intelligence is very generic, so much so that, often it is followed by one or more adjectives that complement and characterize it, we are talking about public Intelligence, Private Intelligence, Military Intelligence, Economic Intelligence, and so on. Intelligence means more things: information, analysis, data collection, news collection, collection of images from one or more sources, but it also means espionage and counterespionage. Sources are also classified on the basis of their reliability, that is, if the source over time has provided information that can be found and useful for those who have requested it and must use it to decide, and they are assigned a level of importance and reliability gradually increasing. In some countries this sector has also developed from the point of view of Economic Intelligence, a necessary tool to preserve a country system as a whole and its business sector, and consequently its position as an actor on the global scene, be it small, medium or large. Development that has also been determined and thanks to methodologies and techniques that have refined and perfected over time, what has changed in particular is the use of modern technologies, and the areas of action, from a purely military intelligence to example during the 2 World Wars, we then came to an Intelligence increasingly cantered on the problems of knowledge of the technological skills of the opponent during the years of the Cold War, then to get to the economic and industrial sphere, and to compete in this field between industrialized and newly industrialized countries, and between countries that fell within the two areas considered.(Stilo Pietro, Economia criminale ed Intelligence, I° Rapporto economia criminale in Calabria).

> When we speak of Economic Intelligence, we mean in fact that process of the activity of the Intelligence cycle, dedicated to receiving and processing news, information, and economic data. Economic Intelligence has a wide field of application, which evolves continuously and branches off in two directions: defensive and offensive. From the defensive point of view, it deals with the protection of a country system as such, understood that is as entire business sector that as individual companies, the protection of trademarks, the protection of intellectual property, the defence of data of public and private activities, the protection of the economy from threats both external and internal, to better allow the development and competitiveness of the Country System. Foreign threats, investments in companies that are strategic for the national economy in fundamental sectors, such as hydrocarbons or advanced technologies, by foreign investors with potentially little transparent objectives may constitute external threats. As regards the offensive area, for example, attention can be paid to those companies (SMEs, but not only) that do not have the structure to transpose, know or recognize activities that can harm their know-how and their ability of insertion and penetration on foreign markets. The addition of the economic element to the Intelligence studies is an element of considerable importance compared to the past, since in recent years the areas of interest have shifted to new sectors, in fact, today the topics of analysis and research of the most advanced Intelligences in the world mainly concern the economy, cyber security and info-war. The first fundamental for the economic and commercial competitiveness of States, the second for the protection of their data in the various strategic sectors, the last to protect themselves from misinformation, that is, from fake news artfully launched by online sites and think tanks to distract, condition and manipulate public opinion or consumers. From this we can deduce why Economic Intelligence is an important discipline, necessary and with many perspectives. (https://www.geopolitica.info/due-nuove-nomine-intelligence-italiana/)

In the near future, but perhaps already today, data control is one of the main forms of power, becoming the most important and indispensable resource of today's and future global context, in this context the use of tools such as artificial intelligence becomes the new frontier that can constitute a new revolution not only in industry, but also in the public administration, as well as in the military and intelligence fields. The new revolution of artificial intelligence has changed and will change more and more in the next twenty years not only the production system, industrial relations and also our daily life, but also the ability to make intelligence, therefore to receive information and data and above all to process these "raw materials". All this together with the diffusion of modern information and communication technologies, also entailed the need to modify and expand the system for collecting news, data and information both for the States, but also for all those subjects that in various capacities operate in the international system, especially multinational companies. Hence also the no longer postponable need for a concrete and growing cooperation between the private and public world, since the exchange of information between the two is now necessary to face new threats and to prevent future scenarios (https://www.agendadigitale.eu/infrastrutture/perche-il-controllo-sui-big-data-cambiera-gli-equilibri-geopolitici-nel-mondo/).

References

Corneli, A.: "Gnosis" n. 7 https://gnosis.aisi.gov.it/sito/Rivista7.nsf/ServNavig/2

Stilo, P.: e AA.VV. "Economia criminale ed Intelligence". Saggio all'interno del I° Rapporto sull'economia criminale della Calabria, della Commissione regionale per l'Emersione del Lavoro non Regolare. ISBN: 9788890911842 (2017)

Stilo, P.: L'Intelligence tra Storia e Leggenda https://www.intelligencelab.org/2017/06/30/lintelligence-storia-leggenda/

Stilo, P.: https://www.geopolitica.info/due-nuove-nomine-intelligence-italiana/ (2019)

Stilo, P.: e Domenico Marino https://www.agendadigitale.eu/infrastrutture/perche-il-controllo-sui-big-data-cambiera-gli-equilibri-geopolitici-nel-mondo/ (2018)

The Future of City/Cities

Domenico Passarelli

Abstract The "smart city" or the "future city" has to balance "new" needs through the widespread and original ICT use (information and communication technology) in some fields such as mobility, environment, energy efficiency and so in urban and local regeneration. A city planned using this approach will have a shared and arranged *governance* where investment will be used for sustainable development respecting intrinsic features of endogenic resources (starting from natural and landscape ones) and unconditionally respecting socioeconomic and environmental potentialities of urban places. Planning a smart city is planning a sustainable system respecting the environment and interpreting social life areas again, connecting new technologies, traditions and the urban places embedded vocation. The city has to be reviewed and replanned. Nowadays we talk about "smart city": systematic planning action where a city is an organized whole and which employs "smart" technologies to make the urban places sustainable from environmental and social point of view. This prototype is efficient thanks to participation and multidisciplinary knowledge.

Keywords Interaction · Regeneration · Intelligent

1 The Contemporary City Natural Perspective

It's important to say in my work the expression "smart city" refers to the contemporary urban setting able to improve its citizens' life quality satisfying their required and performance needs. It's necessary to satisfy qualitative needs no more linked to materialistic and quantitative ones typical of the past rationalist city planning. The "smart city" or the "future city" has to balance "new" needs through the widespread and original ICT use (information and communication technology) in some fields such as mobility, environment, energy efficiency and so in urban and local regeneration. A city planned using this approach will have a shared and arranged *governance* where investment will be used for sustainable development respecting intrinsic features of

D. Passarelli (✉)
Mediterranea University of Reggio Calabria, 89100 Reggio Calabria, Italy
e-mail: domenico.passarelli@unirc.it

D. Marino and M. A. Monaca (eds.), *Economic and Policy Implications of Artificial Intelligence*, Studies in Systems, Decision and Control 288,
https://doi.org/10.1007/978-3-030-45340-4_14

endogenic resources (starting from natural and landscape ones) and unconditionally respecting socioeconomic and environmental potentialities of urban places. Smart city features have to be the fair assignment of human rights and the safeguard and safety of the built and not built setting. This means to value the place physical identity too as expression of an architectural language which featured the history of that place. Nowadays it's necessary to balance physical and intangible aspects through new technologies and re-examining the general urban tools using an original and creative approach. Creativity and technology are successful elements for regenerating a deteriorated city centre and its suburbs. We have to consider the city identity, so each city is smart if it considers its own features related to social and artistic heritage, mobility, school, healthcare, events and so on. Therefore each city will have a typical urban strategy different from the other ones. Planning a smart city is planning a sustainable system respecting the environment and interpreting social life areas again, connecting new technologies, traditions and the urban places embedded vocation. The city has to be reviewed and replanned. Nowadays we talk about "smart city": systematic planning action where a city is an organized whole and which employs "smart" technologies to make the urban places sustainable from environmental and social point of view. This prototype is efficient thanks to participation and multidisciplinary knowledge. It influences the lifestyle change and it needs awareness. It starts a cultural revision because we are surrounded with different types of knowledge as Michel Foucault said.[1]

2 The Ecological and Eco-friendly Conversion as Precondition of the "Smart City"

What are the teenagers and young people from 80 countries of the world, the students and the thousands of city students, to carry out the protest should prompt deep reflection, made dramatic by the data provided by the UN report on the environment, at the Sixth Global Environment Outlook underway in Nairobi, "the damage caused to the planet are so important that, if they will not be taken the emergency measures, the health of populations will be subjected to the increasing threats". But we are also aware that the international community has the scientific knowledge, technology and financial resources required to evolve a model of sustainable development.

To this end the protest action (or proposal) must take a systematic and structured on several fronts but especially must put pressure on the government, involving many forces now engaged in these struggles for ecological conversion of cities and territory (integral recycling of waste, spread of solar, limitation of urban traffic, wastewater reuse, block of concrete, and so on).

The "conversion" is said ecological/eco-friendly because it considers the limits of the environment where we live. These are essentially temporal limits for two reasons: because we are human beings who live in a world that will survive us and so they

[1] Foucault (1969).

concern our deep existence and because they remind us we can't exhaust in a certain time more than the nature can produce and we can't pollute more than the environment is able to regenerate. Increasing demand in order to satisfy wishes, expectations and needs that can't be individually satisfied—that is pointing to what the market offers or doesn't offer nowadays—is a real "social challenge". It's what the contemporary city planning or at least the urban thought of the National Urban Institute suggests asserting a new way to observe places and to understand what people want because their needs are no more materialistic ones but focused on life quality. Energy efficiency and renewable energy, sustainable agriculture, nutrition, sustainable and gentle accessibility mobility, maintenance and repair of the goods, buildings, land, recycling of materials, social regeneration of abandoned and degraded areas, that is all the main themes around which he has been developing, entrenching and specifying the environmental culture in recent decades, today offer real opportunities for their achievement even in limited contexts. To state in general terms the program is easy. Translate it into concrete proposals, it is much more complex. Because demands to compete with the specific characteristics of each territory, both in terms of available resources, both in terms of the needs to be addressed, namely the question of those who inhabit that territory. Just think of renewable sources and energy efficiency—but the same goes for all other areas mentioned—to grasp this point. Each territory has different resources and potential, but also loads and requirements to meet different, how different are the social forces operating there. A unique project, valid for everyone and replicable anywhere there. It's necessary a large activism to make an ecological and eco-friendly reconversion of a factory, an area or society as a whole. We have to involve whole communities with their own associations—if they already exist and if not it's necessary to create them—and at least a part of the local business community and then the local administration or a part of it. Then you have to aggregate/ask/and to build markets for new products; mobilize the common knowledge necessary for the new project; put to work the entrepreneurial skills of those companies that do not see the future in old productions. This is why today it is important to create a "public space" of consultation and participation of those things where we start to discuss city by city, neighborhood by neighborhood, country by country, company by company, allowing everyone to express and to understand their point of view. Working in this way prepares the country to adopt a perspective and projects, or at least of design ideas, viable, including the potential, the actors, the obstacles, the specificities were evaluated.

3 Ecological Conversion is a Matter of Participation

The emergency climate should not be confused with the state of emergency established by law 225/92 for coping with a disaster assigned to civil protection, for limited periods, power derogation from certain laws. Emergency deeper comes from the awareness of the seriousness in which the environment is.

Some things are already clear: the scope and timing of GHG emissions have been agreed in the Paris. But as long as those goals will not be translated into what to do and what not to do are likely to remain on paper.

The objectives should be translated into plans and projects. To articulate these objectives we have to think of us promoting coordination committees and radiating throughout the country environmentally sustainable civic culture. It is a challenge that requires responsibility and planning. The instruments chosen are those useful to found a widespread culture of environmentally friendly and supportive type: Research and development, training and information.

The critical issues that characterize the state of the Italian territories show that both the fight to illegal both to combat the consumption of soil there are no effective measures, and that the available tools are still too otherwise used in our country. In the South, many expansions have taken place without planning, with illegal housing and, consequently, lack the most basic urban facilities. Sometimes they are exposed to significant or damaging cultural and natural heritage risks. The legality in urban and territorial transformations and the responsible use of land should be considered mandatory principles and also practiced throughout Italy. For this you need to confirm our confidence in the integrated and sustainable land use planning reconfirming the task of defining the purpose and behavioral limitations in the use of resources and identify prevention and adaptation actions, priority compared to mitigation measures for the protection of natural resources and the beneficial effects of their good condition on the quality of human and animal life. These conditions are reflected in the government of the territory, on which weighs the absence of national policies and national urban reform, able to guide public action, while exercising at all levels of its own defined principle consistently in different contexts territorial and the interaction between local and national.

4 Regenerate the City Through a Creative and Sustainable Approach

The debate about the "smart city" represents an opportunity to indicate priorities for a new urban development planning process. For town planners it means to change policy, plans and projects suitable for real and potential situation. It's thinking about our cities in the future to get the potentialities of new technologies in harmony with history, traditions, vocation and territory endogenic resources. Cities are a whole of several things: memory, wishes, signs. They are places of exchange, memories, perception, freedom and flavours. So we can share the idea of a city which becomes smart thanks to its innovative requirements (approaching the sustainable development) respecting its own socioeconomic and environmental conditions. Anyway, it's always convincing a theoretical concept that supports what Lefebvre defines a sentimental utopia. Each planner has to consider himself an utopian person because the future city can be built on the dreams of today's one.[2] However, it does not make sense to stop at the concept of intelligent city then if you have polluted neighborhoods, busiest and cemented. The question is then face a number of problems starting from the pressing environmental concerns at a time when it gets stronger the need to promote the efficient use of resources. The essential condition for the regeneration process is satisfactory overall is that it is inserted as part of an integrated strategy, which focuses on resources and potential, making explicit a vision for the city through the definition of medium- and long-term objectives. Integrated strategies for development and sustainable management of cultural and environmental capital may contribute to the overall performance of the cities and rebuild with new systemic balance. The regeneration process should not be reduced to a mere marketing operation in the city to attract businesses and investments, but it should be accompanied by the commitment in solving problems to allow everyone to benefit from regeneration and enjoy a renewed quality of urban life. Hence the need for participation in decision-making increasingly active, which is not limited to a superficial consultation to become an opportunity for real transposition of needs and requirements in the identification of strategies to be pursued in order to govern the processes more consistently and integrated and do not make specific interventions that respond to the logic of profit. The urban regeneration policies and practices in Europe, particularly focused on degraded areas, are significant for the key role of local strategic partnerships, which have contributed to the change of image of the city. Cooperation allowed making substantial investments, with significant results in terms of physical rehabilitation, increase in employment, improving the quality of life in general, although showing a need for a greater focus on the issues of socio-economic disparities, so that all can benefit the changes taking place, and therefore an emphasis on participation. The road ahead is the one of a growing involvement largest in the definition of objectives to be achieved and actions to be taken, despite the path to effective participation is long and winding, but marked by attempts significant, in the direction of an approach more integrated to the resolution of urban issues, with remarkable results in terms of physical rehabilitation, occupational increase, improvement of quality of life in general, although it emerges the need for a greater focus on the issues of socio-economic disparity, so that everyone can benefit from the changes taking place, and therefore a certain emphasis on participation.

All this, in urban planning, is the priority issue in development policies in the coming years. Question to be considered not only as a matter relevant in the planning, but as a policy for sustainable development of cities, limiting urban sprawl and reducing the environmental impacts inherent in the built environment. Urban regeneration is, the opportunity to solve problems such as the lack of identity of a neighborhood, discouraging the use of land not urbanized and focusing on the environmental impact of individual transformation measures, which can not be underestimated if the perspective in which we ask is that of sustainability. The suburbs must no longer be viewed as marginal places of historic, but should be considered urban areas to be integrated in the housing and social fabric of the city. Attention must be paid to the size 'micro',

[2] Amendola (2019).

providing that plans and programs contain even typological design guidelines and to ensure the best use of natural resources and climatic factors, as well as the prevention of environmental risks.

The possibility of carrying out interventions acting on the 'energy flows' affecting the urban organism, it may be an opportunity that can allow environmental improvement even in a perspective of sustainable development. The expected results of these 'local energy policies' will be identified in the realization of a multi-dimensional urban planning that contributes to improved energy and environmental performance of the entire system. In contemporary cities, they are generally deficient elements of sustainability, and this is why we think a new approach to urban planning. By applying the principles of sustainability, starting from the urban form, means taking into account the different dimensions of quality. The most important issues to be taken into account in the planning stages in order to qualify as a sustainable neighborhood are the urban plan; internal mobility; the urban structure; public spaces.

For each of these we can propose some principles and design guidelines specifically aimed at obtaining environmental quality and aesthetic and functional.

The objective to pursue is represented by the construction of urban and environmental policy in the medium-long period in which we disclaim spatially and temporally the interactions between social, economic and environmental spheres that have to do with the urban organism. To articulate the urban planning on the concept of sustainable development and identify environmental strategies is to speak of ecological regeneration of the city and territory.

This determines the depth of the concept of environmental potential as a key indicator of the actual and potential impacts on systems, analyzing the natural cycle of individual environmental factors and their regenerative capacity, identifying the rules and procedures for recovering the imbalances in place and potential. The environment is no longer treated as a sectoral system but is placed within the planning process, introducing priority to the environmental component within the redefinition of the new planning rules for the city. Therefore the idea of sustainability can not be considered a dimension of quality, but it is a principle and a responsibility, that cover every dimension of urban quality. Successes environmental, social quality, Economic and aesthetics can not be achieved if each of these dimensions do not apply in practice criteria and principles of sustainability. This way of planning has totally changed the way we think the use of renewable energy sources, integrating them within the urban structure, sustainable mobility seen as the starting point to create the city not for cars but places where people live mainly standing, green no longer regarded as a decorative element, but urban ecosystem balance factor, thus mitigating the environmental impact of various human activities.

This evolution has been accomplished in numerous European cities, where action has been taken based on multidimensional and integrated approaches. The famous 'Hammarby model' is paradigmatic of such approaches. It represents a district as a closed-loop ecosystem, in which the various waste components in output from the buildings are reused in a virtuous cycle, which only works thanks to the integration of the various parts. The commitment to reduce the environmental burden that characterizes the Hammarby model integrates the aesthetic refinement of morphological and landscape design and attention to the social life of the neighborhood. But not enough to know how to tune the frequency of the city change in theoretical terms, on the transition process that is characterizing the passage of the modern city to contemporary. One city, one today and even more tomorrow's, which no longer refers to the artistic principles theorized by Camillo Sitte,[3] on those of the garden city theorized by Howard or even on Ville radiese of Le Corbusier to name a few but you need to reposition themselves on the most suitable design paradigm to understand and manage systematically ongoing phenomena and interpret them as effectively as possible so that it can fulfill the desirable cities in the future.

For several years we are living a historic step that requires us to consider not only the new variables of a company that is changing rapidly and its geographical new institutional arrangements,[4] Constitutional reforms, but also the ways to build and implement plans and planning programs through new approaches to education that no longer belong to the authoritative and hierarchical planning and/or technical reassuring zoning. The most important change was the case in the light of the relationship with Community policies for the city.

The eleventh Agenda goals that the goal is making cities and inclusive human settlements, secure, resilient and sustainable. Although put particular emphasis on the situation of cities which now account for about 60% of the world population, this goal extends to extend universal access to essential services, sustainable and resilient home in a green environment, the preservation of cultural heritage, protection against disasters. It legitimized the need for expertise governability of system agreement, implementation and project management for the contemporary city. The city is to rethink and redesign. One of the streets indicated today is that of the "Smart City": systematic design interventions they see the city as a coordinated and using "smart" technologies to make sustainable urban centers both from the environmental point of view that from a social perspective. A model that sees the participation and multidisciplinary knowledge of its strength, a model that affects the change in lifestyle, which has need of awareness and you start with a cultural overhaul.

5 The Challenge of the Future

The contemporary city, in its complexity and in the multidimensional interpretation of the constitutive identity characters, necessary and performance (history, current needs and social equity conditions) is the key to sustainability. In a world with an increasing rate of urbanization, the cities directly or indirectly produce almost 80% of greenhouse gas emissions. The problems arising from urbanization are still others: increased poverty, air pollution, large amounts of waste, increased exposure to extreme events (floods to drought), habitat loss and thus loss of biodiversity, reduced soil available for food production and more. This is particularly evident

[3]Sitte (1989).

[4]The law Delrio 56 of 2014 is an example irrefutable.

for the consequences of the most serious threat, that of climate change caused by human activities. The large capacity is to transform the crisis into opportunity factors: and that is the city that you have to look to create a development model based on environmental sustainability and equity. The territorial and urban planning must be able to update their cognitive tools and relationships with dematerialization and networks produced by information technology. The theme of the smart city is a concept borrowed from their own networks needed to distribute and produce energy with renewable sources. The risk is that there are limits to innovation uncritically and without goals: goals that should meet everyone and all together in a synergistic and shared form, in the various meanings of the city of the future: affordable and creative.

6 Artificial Intelligence Tools for City Planning

Urban planning is always eats more than a multidisciplinary approach, involving sectors more and more specific. City planners, therefore, now more than ever become readers of reality, full of new flows, new dynamics, new processes, but especially new technologies. The use and exploitation of the latter allow the absolute knowledge of the city and territory. Everything, therefore, allows the support of decision-making in the new scenarios of the Smart City. The reading of the built comes to be diffused through hyper-tools that allow you to update itself in real time. This causes also the individual today may be, incessantly, receiver element and transmitter of a multiplicity of input/output that affects the outside world.

The interactive and experiential spaces that respond and adapt to different emotions, return "humanity" of things, and, above all, show an intelligent use of which resource technology that not only improves and intensifies itself but helps to provide reading suggestions (of objects, artifacts, urban landscape,…) and services not complied, unexpected and embracing—through, of course, new ways to interface with reality, today represent significant elements in the processes of sustainable development and integrated land, in 'implementation of marketing strategies aimed to the spread of knowledge and the promotion of places, in the activation of intelligent communities (smart city and smart community). In his book Me++, Mitchell (2004), verifying the effects produced by digital information in different disciplinary fields and about features unrelated to it in the past, states the binary code is the most powerful tool we have to express our planning ability and to make it a real action. "We are becoming real inhabitants of electronically mediated settings rather than mere users of computational tools,[5] as Ratti and Claudel show through recent experiences in the Senseable City Lab. According to the authors, talking about the development of the "city" idea, there is a turning point defined by the intersection of technology power and built setting. So we can definitely agree with Mitchell when he states: "the net is the urban site that faces us, a call for planning and building the bits city, as

[5]Mitchell (1999).

years ago a narrow peninsula next to Meandro became Mileto site".[6] The attention is focused on different structural criteria through which this tool allows to organize communication and to enjoy the city and its territory. So it's possible to stress its worth is in the type and in the quality of its relationships with the context. Therefore Membrana smart isn't an object in the space but a tool to create the space, a part of a city set-up and of the territory where it is. Technological advances have enabled the advancement of artificial intelligence especially for the management of the large amount and variety of informative data that allows the simultaneous management of multiple amounts of data.

In this sense, it is essential to go on the perfect knowledge of the building that does not become finer information to itself, but guidance tools of static and dynamic decisions. The relief with sophisticated equipment such as laser scanners or unmanned aircraft (drones) lets you scan the reality immediately and with pinpoint accuracy. The data collected can also be used transversely through the new communication technologies "furniture", able to operate in the context of web-based information systems. The knowledge of the area then passes the virtual prototype made of 3D reconstructions of parameterized objects to the true objectivity of places and the result of the acquisition of real territories. It all leads back to the new way of conceiving the territory and consequently the new way of conceiving the project action to any reference scale, designed with "transcalare approach". The artificial intelligence that allows you to manipulate a given quality in the mapping of information, which exceeds the third dimension allows you to: (a) protect the territory; (b) enhance and promote the cultural and environmental heritage; (c) promote and market the tourist offer of the place in order, also, to improve their economic and social condition and therefore, more generally, the quality of community life; provide a product/system for the use of a "Creative Culture" without perceptual and sensory barriers (and architectural).

References

Amendola, G.: Looks in the Modern City. Dedalo, Bari (2019)
Foucault, M.: Archeology of Knowledge. Rizzoli, Milan (1969)
Mitchell, W.J.: E-topia. MIT Press, Cambridge (1999)
Mitchell, W.J.: The City of Bits. Spaces, Places and Information Highways. Electa, Milan (1997)
Mitchell, W.J.: Me ++, Cambridge, Mit Press (2004)
Sitte, C.: The Art of Building the City (1989) (the original edition)

[6]Mitchell (1997).

The manufacturer's authorised representative in the EU is Springer Nature Customer Service Centre GmbH, Europaplatz 3, 69115 Heidelberg, Germany. If you have any concerns regarding our products, please contact ProductSafety@springernature.com

Printed and bound by CPI Group (UK) Ltd, Croydon, CR0 4YY
30/09/2026
02225607-0017